NATURE'S BEST HOPE

NATURE'S BEST HOPE

A New Approach to Conservation That Starts in Your Yard

DOUGLAS W. TALLAMY

TIMBER PRESS | PORTLAND, OREGON

Published in 2019 by Timber Press, Inc.

The Haseltine Building
133 S.W. Second Avenue, Suite 450
Portland, Oregon 97204-3527
timberpress.com

Printed in China

Text design by Laura Shaw Design
Jacket design by Adrianna Sutton

ISBN 978-1-60469-900-5

Catalog records for this book are available from
the Library of Congress and the British Library.

CONTENTS

Introduction

IN 1903, WITH THE STATE OF ARIZONA on the verge of mining the Grand Canyon, President Theodore Roosevelt stood on the canyon's lip, gazed out over its unique magnificence, and uttered the five words that would save it: "Leave it as it is." Unfortunately, because only 5 percent of the land in the lower forty-eight United States is now in anything close to a pristine, self-sustaining ecological condition, we've lost the opportunity to save most of our country from such development. Ninety-five percent of the country has been logged, tilled, drained, grazed, paved, or otherwise developed. Our rivers have been straightened and dammed (damned?), and several no longer reach the sea. Our air has been polluted, our aquifers pumped nearly dry, and our climate changed for centuries to come. We have purposefully imported thousands of species of plants, insects, and diseases from other lands, which have decimated many native plant communities on which local food webs depend, and we have carved the natural world into tiny remnants, each too small and too isolated to support the variety of species required to sustain the ecosystems that support us.

I could go on, but this is not a book about the pox we have delivered upon the environment and thus upon all of our houses. It is a book about a cure for that pox—a cure that will require small efforts by many people but that will deliver enormous physical, psychological, and environmental benefits to all. This may sound like hyperbole, but we have learned that restoring the natural world benefits not only other species, but *Homo sapiens* as well, in ways no one imagined only a few years ago.

Despite the current political climate, I believe we are on the cusp of a new environmental ethic, one that will (must) be adopted, not just in "blue states," but in "red states" as well; not just in the United States, but worldwide; not just by tree-hugging environmentalists, but by everyone. It is quite possible that historians will call the coming decades "The Age of Ecological Enlightenment." I am not a soothsayer or a visionary; I am an ecologist who makes this claim with confidence, because it is the only option left for *Homo sapiens* if we want to remain viable in the future. In his 1949 conservation classic, *A Sand County Almanac*, Aldo Leopold wrote, "There are some who can live without wild things and some who cannot." He was referring to our emotional connection to the natural world, a connection that fewer and fewer people seem to have these days. The message I have tried to convey in this book is that, whether we like nature or not, none of us will be able to live for long in a world without it.

I have found that most people fall into one of three groups: they like plants, they like animals, or they like neither. I have attempted to address all three groups in this book. My goal is to convince people that we will lose most of our plants if we lose most of our animals, and we will lose all of those animals if we don't take care of our plants. For people who don't care for or about the flora and fauna around them, the hardest group of all to engage, I have also done my best to explain why we will lose humans if we don't preserve the plants and animals that keep our ecosystems healthy and sustaining.

We haven't always thought about our individual impacts on the local environment, and consequently those impacts have been negative more often than not. Today, however, we must develop a new mindfulness about how our everyday actions affect the physical and living world around us. In this book I argue that we can no longer tolerate actions that degrade our local environment; there are simply too many of us for the earth to sustain the cumulative impact. A neutral impact is not good enough, either, for we no longer have the option of leaving things in their current degraded state. We must now act collectively to put our ecosystems back together again. What is needed is nothing less than a cultural transformation: rather than acting as if we were independent of nature, we need to behave a little more reverently or respectfully toward nature, as if we were the product and beneficiary of a vibrant natural world, rather than its master.

Today most people live in what I call the great suburban/urban matrix, and we hardly interact with the natural world. Unfortunately, our ignorance of nature has led to a dangerous indifference about its fate. The local disappearance of once-common plants and animals does not bother us because we have grown up with no knowledge of these species, and we cannot imagine why they are important to us. We do not teach our children that plants and animals actually generate the life support systems we all require. Plants produce our oxygen, clean our water, and delay its journey to the salty sea. They store atmospheric carbon that would otherwise wreak even more havoc with our climate. Plants build our topsoil and hold it in place, and they prevent floods when we leave enough of them in our landscapes. Animals, in turn, provide pest control services and pollinate not just our crops but nearly 90 percent of all plant species. We are living off of the ecological interest that was generated by a healthy ecological bank account long ago, but we are eating up the principal of that account at a steady and alarming rate. Agriculture lands, cities, and the vast suburban tracts that surround them have replaced natural areas in so many places that not enough nature remains to generate the natural capital on which our lives depend.

Gardening is like cooking. It is tempting to cook only with the goal of achieving great taste, with no thought of healthy eating, but that often results in tasty concoctions so full of fat, sugar, and salt that they are deadly in the long run. Similarly, it is tempting to garden only for beauty, without regard to the many ecological roles our landscapes must perform. All too often, such narrow gardening goals result in a landscape so low in ecological function that it drains the vitality from the surrounding ecosystem.

Think for a moment about your own yard. If you are a typical homeowner east of the Mississippi River, about 90 percent of your landscape is lawn, and your yard contains only 10 percent of the tree biomass that it supported before your house was built. When choosing plants for your landscape, you considered only their decorative value, and you chose the same few species that your neighbors and their neighbors chose. You gave no thought to the many roles your plants could play within your local ecosystem if they were contributing members of that ecosystem. So 80 percent of the plants in your yard are species that evolved in Asia, Europe, or South America—species that are unable to support the complex food webs necessary to sustain ecosystem function in your area. When you look out your window, nothing moves. This does not bother you because you grew up in a house with a yard in which nothing moved; you think a yard with no animal life is normal—after all, animals belong in nature, and nature is someplace else. What's more, your civic or homeowner association has passed rules suggesting that building landscapes that do not support wildlife is good land stewardship.

Now put on your ecological thinking cap. If your property does not generate all of the ecosystem services you and your family need to live well, you will have to borrow services that were generated somewhere else. Chances are your neighbor's yard looks just like yours and is just as biologically depauperate (lacking in numbers or varieties of species), so you will not be borrowing ecosystem services from your neighbor. Your township's open spaces have most likely been converted to soccer and baseball fields or vast expanses of lawn with paved tracks encircling them for joggers and dog walkers, so you will not be borrowing ecosystem services from your public spaces either. In the past, residents lived off the services generated from the diverse ecosystems that surrounded a town or city, but those ecosystems have shrunk in size, diversity, and effectiveness by the day. So where will you get what you absolutely need to live in the coming years?

Many people are surprised to learn that former President Richard Nixon understood the limits to the amount of abuse our natural resources could endure. In his 1970 State of the Union address, he said, "We can no longer afford to consider air and water common property, free to be abused by anyone without regard to the consequences. Instead, we should begin now to treat them as scarce resources, which we are no more free to contaminate than we are free to throw garbage into our neighbor's yard." In this book I will argue that this sentiment now applies not just to human waste, but to ecological contamination as well.

Just as we are not free to throw garbage into our neighbor's yard, we are no longer free to release countless propagules of invasive species onto our neighbor's landscape. We are no longer free to flood our neighbors with stormwater that our huge lawns cannot absorb; nor are we free to deplete our neighbor's aquifer by watering our thirsty grass. None of us has the right to destroy the diversity of life that once thrived on our properties—life that is required to run the ecosystems that keep us and our neighbors alive. We do not have the right to starve local pollinator species by removing the native flowers on which they depend. We do not have the right to heat up our neighbor's airspace by cutting down the trees on our property, nor do we have the right to change our neighbor's climate by pumping carbon dioxide into the air when we mow our lawns. In short, we no longer have the right to ignore the stewardship responsibilities attached to land ownership. Our privately owned land and the ecosystems upon it are essential to everyone's well-being, not just our own. Abusing land anywhere has negative ramifications for people everywhere. UNESCO, the United Nations Education, Scientific and Cultural Organization, designates biosphere reserves as places of ecological significance. But clearly this is misleading, for all places have ecological significance, not just the few places we have not yet destroyed. In short, gardening in the traditional sense is optional, but earth stewardship is not.

Nixon's vision for a sustainable relationship with the natural world was remarkable for its clarity and wisdom, and its ecological implications went far beyond the obvious need to protect our air and water. Nevertheless, the notion that the earth and its natural resources are not infinite and must be managed wisely (by everyone) for the greater good must have been ahead of its time, because it was largely ignored. In many ways, our cultural relationship with the environment has improved little since the 1970s, and there is a growing sense of alarm among the general public, even if most of us still struggle to articulate it. Lucky for us, the ecological Armageddon we have been so blindly encouraging is largely reversible if we simply adopt a new relationship with nature. We can save the natural world—and ourselves, for we are part of it and it is an inextricable and essential part of us—if we stop segregating ourselves from nature and learn to live as a part of it. We must shrug off our age-old adversarial relationship with nature, the "nature versus us" attitude that may have worked for our ancestors but is deadly to us now. We must be wary, very wary, of claims that all of our needs in the future will be met by technology, that we no longer require natural systems and the interrelationships that they comprise, and that the loss of other species is not only inevitable, but a good thing because it signals the arrival of the Anthropocene, an age when all of the earth's resources will be usurped for human needs. Finally, we must accept the new reality that how each one of us treats our local biological heritage impacts not just ourselves but our entire community.

As far as we know, the only complex life forms anywhere in the universe, and certainly the only ones you and I will ever interact with, occur in a thin film surrounding the earth—the biosphere. We have boldly assigned ownership to this unique combination of water, organic molecules, and biologically favorable weather: Tom owns this part, Dick owns that part, Harry owns the part over there, and Mary owns the parcel down the street. So be it; but along with this ownership comes the responsibility of stewarding the only known life in the universe, which is perhaps the most awesome responsibility of all. In the coming pages, I have tried to personalize our conservation challenges and our responsibility to meet them. One of my central messages is that effective conservation is not beyond the reach of the individual; indeed, it is your efforts as an individual that will determine whether we succeed or fail, and whether we live in a world thriving with life or in one in which little stirs.

In short, this book is about fixing problems. The good news is that we can fix our ecological problems by indulging rather than sacrificing. It has been very difficult to address environmental issues by asking people to give up something—their suv, clothes dryer, sirloin steak, or the idea of having a third, fourth, or fifth child—in order to gain the long-term benefits from good health, a moderate climate, clean air and water, and life-sustaining ecosystems. These are certainly real benefits; they are essential, in fact, to a modest quality of life. But even if we behave well today we won't realize these benefits until sometime down the road, around life's next corner, in that nebulous uncertainty we call the future.

And therein lies the problem. Humans are not genetically programmed to care about the future. Yes, we care about tomorrow, but not as much as today. Next week? Maybe, but next year? Ten years from now? Thirty years from now? No way. Throughout our evolutionary history, those of us who worried about meeting our immediate needs were more successful than those who planned and allocated resources for a future that we often never realized. So in this book I will not be asking for sacrifices that will build a better world later on. I will suggest actions that heal our damaged landscapes right now, actions that create immediate, short-term gains for humans. That such actions will also deliver long-term ecological benefits is just icing on the cake, as important as that icing is. In the world I envision, landscaping practices will no longer degrade local ecosystems; landscaping will become synonymous with ecological restoration. We will not be living with less; we will be enriching our lives with more—more pollination services; more free pest control; more carbon safely tucked away in the soil; more rainwater held on and within land for our use in a clean and fresh state; more bluebirds, orioles, and pileated woodpeckers in our yards; more swallowtails and monarchs sipping nectar from our flowers. Indeed, more species of all kinds will inhabit our landscapes, increasing the stability and productivity of our ecosystems. This proactive approach to earth stewardship will no longer be the unfulfilled dream of a few environmentalists, but a culturally embraced imperative, not only because we have no other choice, but because it works. It is nature's, and thus humanity's, best hope.

CHAPTER ONE

The Dreamers

Conservation biology . . . [is] a discipline with a deadline.

—E. O. WILSON

MANY VISIONARIES have recognized that humans do not have a sustainable relationship with the natural world that supports them, and they have worked or continue to work to improve that relationship. Conservation pioneers who immediately come to mind are George Perkins Marsh, Theodore Roosevelt, and John Muir, who helped create our system of national parks, monuments, and preserves; those who organized the public sector of our national environmental movement, such as Edward Abbey and David Brower; courageous people such as Rachel Carson and James Hansen, who employed modern science to expose environmental threats from big business; science writers such as Bill McKibben and Elizabeth Kolbert, who have dedicated their careers to bringing critical conservation messages to the public; and countless others who have practiced conservation successfully at the state and local levels, but who will never receive the recognition they deserve. Among all of these dedicated people are two of the most respected giants in the world of conservation: Aldo Leopold and

Edward O. Wilson. They stand above the rest, particularly for me, because the many unique insights in their writings stimulated my own ideas and inspired me to write this book.

ALDO LEOPOLD

On 11 January 1887, in Burlington, Iowa, a small town on the banks of the Mississippi River, Rand Aldo Leopold entered the world he would come to love so passionately (Meine 2010). He was born with an innate interest in all things wild. Although he was guided by his parents, he needed little encouragement to become a naturalist, forester, wildlife biologist, professor, restoration ecologist, author, philosopher, and perhaps the most influential conservationist of the twentieth century.

Aldo Leopold was fascinated with the natural world; wherever he traveled, he habitually recorded the seasonal cycles of the plants, birds, and other animals he saw, as well as the climate cycles that influenced them. In fact, what distinguished Leopold from other naturalists of his day was his interest in how members of a community interacted with one another and the physical world they inhabited. Today we would call Leopold a systems biologist; he was well ahead of his time, and he recognized and was intrigued by the complexities of ecosystems decades before the term was even coined.

But like Teddy Roosevelt before him, Aldo Leopold's early relationship with nature was somewhat schizophrenic. In Leopold's day, the most common introduction to wild things came through hunting, and that is how his father first exposed him. Not long before young Aldo took his first foray into the woods, hunting was considered more a necessity than a sport, and the notion that wildlife existed to be hunted was deeply rooted in his culture. The fledgling field of wildlife management consisted of shooting as many wolves, cougars, and bears as possible to encourage growth of populations of the deer, moose, and elk that hunters pursued. After all, predators were thought to compete with the needs of hunters and therefore ought to be eliminated. One of the primary responsibilities in his first job as a forester in the U.S. Southwest was to kill as many predators as he could—and he was good at it. Perhaps the most ironic aspect of Leopold's life was that it was he who shot and killed one of the very last Mexican gray wolves in the United States.

Aldo Leopold is often considered the father of modern conservation.

But it was also Leopold who first recognized the results of such carnage. In what would much later be termed "trophic cascades" (Hairston et al. 1960), Leopold saw that when top predators were eliminated from an ecosystem, the herbivore populations they once kept in check exploded (despite hunting pressure) and the vegetation that supported the entire ecosystem became disastrously overbrowsed. This, in turn, led to starvation and disease for the very species predator removal was supposed to be helping, as well as for myriad other species that depended on healthy plant communities. In a stark departure from accepted wildlife management protocol, Leopold suggested that members of the top trophic level—the predators—were essential to the well-being of the trophic levels beneath them—the smaller predators, herbivores, and particularly the first trophic level, the plants that fed them all. Removing wolves, cougars, and bears created an imbalance in the energy flow through an ecosystem that cascaded down to the plants, and the ecosystem collapsed to a paltry remnant of its former abundance and diversity.

Some hunters still protest when Leopold's advice is heeded and top predators are protected, but nearly a century of research has proven him correct. From starfish in tidal pools of the Pacific Northwest, to sea otters off the California Coast, to the overabundance of white-tailed deer in the predator-free East, as well as dozens of other examples from around the globe, studies have shown that top predators are not simply desirable members of a community but are essential to the sustainability of their ecosystem.

The far-reaching impacts of removing top predators from an ecosystem came into focus more clearly than ever before when wolves were returned to the Yellowstone ecosystem in 1995, some seventy years after they had been exterminated (Stolzenburg 2008). In just a few years, Yellowstone's wolves reduced the moose and elk populations, eliminating overbrowsing by the herbivores, and caused a truly remarkable recovery of other species, including the bison with which exploding populations of moose and elk had been competing; beavers and all of the species associated with the stream and wetland ecosystems they create; grizzly bears, bald eagles, and ravens that depend on wolf kills for scavenging; six species of song birds that breed in restored streamside vegetation; and, of course, the all-important willow, aspen, and cottonwood populations upon which all of these species depend. Never had Aldo's foresight been so dramatically demonstrated, but it was forty-seven years too late for him to enjoy himself.

Leopold's recognition of the importance of predators and the trophic cascades that result from their removal was only his first contribution toward tempering humanity's relationship with the natural world. In 1924 he was transferred from New Mexico to Madison, Wisconsin, where he soon joined the faculty at the University of Wisconsin, but not before he had written a management plan for New Mexico's Gila National Forest, which that year had become the country's first official wilderness area. He would flourish at Wisconsin, becoming the first chair of a new program in game management, writing the first and arguably most famous textbook on wildlife management, and founding The Wilderness Society. Despite his successes, he was deeply disturbed by what he saw in the environment around him.

In almost every way, people were destroying the natural world and the coevolved associations that glued it together. Society's relationship with what he famously called "the land" was not a relationship at all, but a unidirectional exploitation of resources that returned nothing. Farmers overworked the soil in ways that encouraged catastrophic erosion, grasslands were severely over-grazed nearly everywhere, rivers were treated as sewage receptacles and garbage dumps, marshes and prairie potholes were drained or filled, and what remained of virgin grasslands was plowed under. Repeated clearcutting and burning transformed majestic forests into wastelands, and wildlife was slaughtered so often and in such numbers that many species were extirpated from Wisconsin. The sandhill crane, which Leopold revered as the symbol of the untamed past, was hunted relentlessly, and by the time he moved to Wisconsin, only a few cranes remained in the far north.

But Aldo Leopold had a dream. He dreamt of a time when people humbly accepted their roles as citizens of the natural world rather than its conquer-ors, a time when the land was not viewed as a commodity to be exploited but as the source of our continued existence. He longed for a time when people appreciated and even respected wilderness, not just as a hunting or recreational playground, but as a truly awesome and unimaginably complex machine that required all of its parts to function well.

These ideas, these hopes, and these dreams didn't come to Leopold over-night; they came from a lifetime of thoughtful observation, reflection, and informal experimentation. In 1935, his family bought a degraded tract of land in the Central Sand Plains of eastern Wisconsin. Once a poorly managed farm, the eighty acres were barren scrub that supported little life when the family bought it. Leopold built a small summer home he fondly called "the shack," and for the next thirteen years the family restored the ecological integrity of their tiny piece of the world. Through trial and error, they learned how to bring life back to their land by rebuilding prairie, savannah, and marshland where it had once been. Leopold painstakingly recorded the rapid return of the wild things he loved and was so encouraged by the success of the restoration that he wrote his masterpiece, *A Sand County Almanac*, with hopes of inspiring a new land ethic that would transform how people viewed and interacted with nature (1949). He viewed conservation as a state of harmony between people and land and foreshadowed the concept of environmental sustainability when he stated, "A thing is right when it tends to preserve the integrity, stability, and beauty of the biotic community. It is wrong when it tends otherwise." On Easter Sunday 1969, twenty-one years after Leopold died of an untimely heart attack, sandhill cranes returned to the marshes of his property, an event that still brings tears to my eyes each time I think about it.

Remarkably, *A Sand County Almanac* was rejected by several publishers before being accepted for publication by Oxford University Press the week Leopold died. Though it sold slowly at first, it eventually became wildly popular, and today more than two million copies have been printed in fourteen languages. Most people agree with Leopold that we need to adopt a land ethic that respects and protects all members and aspects of nature in harmony with the needs of people. The Aldo Leopold Foundation, founded in 1982 by his wife, Estelle, and their five children, fosters his concept of the land ethic through education and a celebration of the man himself. Yet, as I look around, I wonder where I see this land ethic in practice? Oh yes, there have been great strides in the environmental movement since Leopold died in 1948, and much of the credit for these necessary changes goes to him. Powerful organizations such as The Nature Conservancy, Sierra Club, National Audubon Society, as well as smaller land conservancies around the country, have protected many wild places beyond our national parks and wilderness areas. In the United States, legislation in the 1970s, such as the Clean Air Act and Clean Water Act, have noticeably curbed point source pollution, while the Endangered Species Act created a national recognition that extinction at the hand of humans is not OK.

Nevertheless, we cling to the notion that nature should be saved where nature remains, not where humans work, live, farm, or play. Though persuasive and moving, Aldo's plea for a land ethic has thus far been unable to change the nearly universal belief that people are here and nature is somewhere else. And this is where philosophical musings about conservation have run head-on into the brick wall of the earth's finite size and resources. The ecosphere, the frighteningly thin zone at the earth's surface to which life is constrained, is not getting any bigger. There is no more land today than there was 600,000 years ago when *Homo erectus* first harnessed fire. In fact, the resources that support life on earth are all under pressure from growing human populations and consumption. And yet we continue to grow, continue to build, continue to sprawl. Where is our expression of an ethical relationship with the land and the life it supports when we fragment forests to add another housing development, pave more roads, seed a new sterile lawn, build another shopping mall, or expand another airport? It is not part of the discussion. After all, we are told that we need development, and that our economy must continue to grow forever, even though such growth is antithetical to the laws of physics. Conservation is fine as long as we do it in ways that do not constrain the human activities we call progress—as long as we do it someplace else. And so, we continue to exile the natural world to increasingly small parcels of land where no people reside. We admire Leopold's concept of a land ethic, but sadly, we have yet to apply it in a meaningful way.

EDWARD O. WILSON

In 1929, while Aldo Leopold was busy expanding the wildlife management program at the University of Wisconsin, Edward Osborne Wilson was born in Birmingham, Alabama. Wilson split his early years between Mobile, Alabama, and Washington, D.C., and, like Leopold, he was irresistibly drawn to the natural world. His route to becoming the world's foremost myrmecologist (ant specialist) was a circuitous one, shaped largely by mishap and serendipity. Like so many boys, E. O., as he would fondly become known decades later, was fascinated by snakes, both poisonous and otherwise. An unfortunate encounter with one of his poisonous friends convinced him to spend more time with less aggressive creatures, so he added fishing to his activities. One day, while fishing alone off Paradise Beach, Florida, one of his favorite haunts, he jerked a hooked fish out of the water so quickly that it hit him in the face, and one of the spines on the fish's dorsal fin pierced his right eye. One can only imagine how painful that must have been, but he was reluctant to give up his afternoon of fishing, so he didn't seek medical treatment at the time. Neither did he do so after he arrived home. After several months, a thick cataract developed, which led to the surgical removal of his lens.

Wilson's fishing accident and blinded right eye was traumatic, to say the least, but it did have a decided bright spot: Wilson had superior 20/10 acuity in his left eye, and he discovered that he could see tiny things exceptionally well. This ability quickly led to a fascination with insects, and though he chased butterflies and dragonflies like all young entomologists, he decided he liked flies best of all. Wilson's passion for flies did not last long, however, because flies needed to be mounted on pins for study, and insect pins were scarce during the lean years of World War II. For some reason, small glass vials were not, so in a pragmatic decision, Wilson shifted his focus from flies to ants that could be stored by the dozens in vials. With the encouragement of experts at the Smithsonian's National Museum of Natural History in Washington, he began a detailed account of the 136 species of ants of Alabama. At the age of eighteen, Wilson published his survey, which included the first account of a colony of dreaded fire ants in the United States.

By any measure, Wilson has had an exceptional career; while most scientists are lucky if they make a single seminal contribution to their field, it is difficult to keep track of

Edward O. Wilson is one of the most eminent biologists of our time.

the many ways and times he has substantially advanced science over the course of his long career. In 1956, he coauthored with William Brown the article "Character Displacement" in the journal *Systematic Biology*, which was to become one of the most frequently cited scientific papers of all time. In the lab, Wilson was the first to demonstrate the all-important roles of pheromones in social insects, a discovery that not only explained how insect societies worked, but one that became the basis of thousands of future research papers by dozens of top scientists. Collaborating with Robert MacArthur, Wilson developed the highly influential equilibrium theory of island biogeography, the stimulus for hundreds more research papers. He also created, defended, and brought to fore the field of sociobiology, which, itself, spawned the new discipline of evolutionary psychology. In his spare time, Wilson has been an uncommonly prolific and accomplished author (to date, he has written twenty-nine books) and is the only scientist to have won the Pulitzer Prize for General Nonfiction twice. I include Wilson here, however, because of his tireless role in promoting the conservation of biodiversity.

First coined in 1968 by Raymond F. Dasmann and introduced to the scientific community by Thomas Lovejoy more than a decade later, the term "biodiversity" (a contraction of biological diversity) means exactly what it implies:

the diversity of life forms on Earth (Lovejoy 1980). Included in the definition is not only the vast number of different species, but also the variation of their genes and the diversity found within populations of those species that is so necessary for their adaptability, ongoing evolution, and thus continued existence in a rapidly changing world. Many scientists have extended the definition to include the diversity of ecosystems and biomes as well.

In 1992, Wilson wrote *The Diversity of Life*, the first book to describe the seemingly infinite diversity of species that reside on Earth today, as well as the many imminent threats to their future. If, as I have claimed for some time, that knowledge generates interest, and interest generates compassion, then Wilson has compassion to spare for life on Earth. And if compassion generates action, Wilson has also been exemplary. His primary tool has been his pen, and he has written time and again about the need to preserve biodiversity. He followed *The Diversity of Life* with seven more books on this theme: *The Biophilia Hypothesis* (1993), *In Search of Nature* (1996), *The Future of Life* (2002), *The Creation: An Appeal to Save Life on Earth* (2006), *A Window on Eternity: A Biologists Walk Through the Gorongosa National Park* (2014), *The Meaning of Human Existence* (2014), and his latest book at this writing, *Half Earth: Our Planet's Fight for Life* (2016).

In *Half Earth*, the normally eloquent Pulitzer Prize winner does not mince words. Time is running short, he says. So is clean water, fresh air, the bounty of the sea, the area of remaining rainforests, and the numbers of plants, reptiles, amphibians, fish, mammals, and birds—all of which are essential to sustaining the living portions of our planet. Wilson claims that emergency measures are required to stabilize the biosphere before the sixth great extinction—the sixth time in world history that a large number of species has disappeared in rapid succession—renders recovery impossible: we must set aside half of planet Earth as a human-free natural reserve to preserve biodiversity.

Say, what? Save half of the earth? Could he be serious? Indeed, he could. Using arguments from species area curves and his own theory of island biogeography, Wilson describes how saving half of the earth could stabilize 80 percent of its species. (Michael Rosenzweig [2003] writes that he is not as optimistic; his own analyses of what generates and sustains biodiversity predict that preserving half of the earth would, in the long term, save only half of its species.) Wilson claims that we cannot select a random half; we have always been generous in setting aside areas such as mountain ridges, the driest deserts, and tundra that are incapable of supporting humans, or most other species. No, to save a lot of species effectively, we must save areas that best support those species, such as tropical forests (both wet and dry forests), much of the African plains, scrublands of southwest Australia, important parts of all major biomes, and half of the fishable ocean. Echoing Leopold's mantra, Wilson reminds us that "the biosphere does not belong to us; we belong to it" (2017). If we continue to ignore the health of the earth, we are dooming ourselves.

To many, including some of the most zealous conservationists, Wilson's *Half Earth* manifesto seems as preposterous as it is noble. Though he makes an irrefutable case for the need to save viable populations of the earth's biodiversity, how can we possibly put aside half of the earth when nearly eight billion humans already occupy most of it? Protecting our oceans should be relatively easy, and there are still substantial chunks of unprotected tropical forest in the Amazon and Congo River basins that can and must be preserved. But what about the rest of the land? We already intensively farm or graze nearly half of the earth's land surfaces (Owen 2005). The remaining 50 percent is divided between cities, suburbs, vast complexes of infrastructure, the patchwork of fragmented second-growth habitat scattered here and there, uninhabitable areas, and the areas already preserved, which total only 17 percent of the earth's land surface.

Biodiversity is not equally distributed across the earth and, at least in the United States, it is not correlated with the land that is preserved. Most of our nation's biodiversity is found east of the Mississippi River, while most of our protected areas are west of the Mississippi (Jenkins et al. 2015). Protecting what is not needed for agriculture, in the sense that most people interpret Wilson's mandate—that is, creating preserves in the traditional model of our national parks or wilderness areas—seems impossible, because it *is* impossible with the current human population size.

Fortunately, there is an alternative.

DREAMS TO REALITY

Giving up is not an option; our current model of destroying the biosphere to expand the human footprint is not now and never has been sustainable. We need a new conservation plan, one that sustains the living systems we depend on everywhere, where humans dwell as well as where they do not. We must abandon our age-old notion that humans and nature cannot mix, that humans are here and nature is somewhere else. Starting now, we must learn how to coexist. When Leopold moved east from New Mexico, he recognized that the conservation model he had followed in the West, setting aside large tracts of government land, would not work in Wisconsin, because most land there was in privately owned farms and ranches. His solution was to teach farmers and ranchers techniques to restore and conserve the natural resources on their own lands. With incredible foresight, Leopold suggested "rewarding the private landowner who conserves the public interest."

Today, with more than 83 percent of the United States privately owned and 86 percent east of the Mississippi River in private hands, it is clear that Leopold's approach is an important part of the solution. If conservation is to happen, it must happen largely on private property, but not just on farms and ranches; it must include all types of private property, from the smallest city lot to the largest corporate landscape. This is not to diminish the critical conservation role public preserves now play; they are the repositories of the species we want to save as well as the genetic diversity that will enable us to save them. But public preserves are not enough to sustain biodiversity into the future. The United States could become a model for the rest of the world in this regard. If we can save biodiversity here, where aggressive economic development has been the goal for centuries, where McMansions have replaced modest homes in affluent communities across America, where we have paved over an area larger than the state of Ohio, where we have built airports twice the size of Manhattan, where mega-farming in the absence of hedgerows was invented, and where biological wastelands we call lawns are core symbols of wealth and status and now occupy a space the size of New England, we can save biodiversity everywhere.

Our relationship with the earth is broken. Leopold and Wilson offered ways to fix it, but the conservation approaches developed in the twentieth century are not inclusive enough to realize their dreams. We need a new conservation toolbox, packed with more expansive tools. New knowledge will be our most important tool, followed by a cultural recognition that conservation is everyone's responsibility—not just those few who make it their profession. Every day we are learning more about how to redesign both public and private landscapes in ways that meet the aesthetic, cultural, and practical needs of humans without devastating the resources needed by humans and other species. We are learning how to convert at least half of the area now in lawn to attractive landscapes packed from the ground to the canopy with plants that will sustain complex food webs, store carbon, manage our watersheds, rebuild our soils, and support a diversity of pollinators and natural enemies. In other words, we are learning how to create landscapes that contribute to rather than degrade local ecosystem function. Finally, we are learning how rapidly the animals return to our yards, parks, open spaces, neighborhoods, and even cities when we landscape sustainably.

These are exciting times. The necessary task of restoring ecological function to the land lies mostly before us. But it is an exhilarating, entertaining, and hugely rewarding task. Leopold once lamented that "the oldest task in human history" is "to live on a piece of land without spoiling it." In the past, we have not known how to do this, but now we do know how. There are few of us who cannot improve our relationship with the land we own. Most of us bought land that was already spoiled, and we must now fix it. Leopold and Wilson are not the only people who have dreamt of preserving the wonders of the natural world. Their dream has been shared by millions of us mere mortals. In this book, I hope to tap the energy of the dreamers among us and show how to make many, if not all, of these dreams come true.

CHAPTER TWO

A New Approach
to Conservation

We can't solve problems by using the same kind of thinking
we used when we created them.

—ALBERT EINSTEIN

CONSERVATION BIOLOGY as a science is young, but it has existed as a philosophy since the Middle Ages (Liddiard 2007). The first serious efforts to protect natural areas from overexploitation were enacted some 500 years ago by European aristocracy as a means of protecting their favorite pastime: hunting. As the great forests of Europe began to disappear, kings and feudal lords realized that to have plenty of game to hunt for sport, they needed to protect the animals and the forests that housed them from peasants who needed food and fuel.

Conservation by popular demand did not take root in Europe until the 1800s, when British artists started to change the subjects of their paintings from human forms and religious events to the beauty of the natural world. Celebrating the positive aspects of the natural world was a substantial philosophical shift in Western cultures at that time; before then, nature was viewed not as something with inherent value and beauty but as something that was dangerous and scary. As Europe was tamed, however, the fear of being eaten by a wolf or a bear was gradually replaced by the sense of well-being associated with peaceful pastoral settings. When Europeans came to North America, the vastness of its wild places was intimidating and once again viewed as something to be subdued. Wilderness could not be farmed and settled until it was conquered. But the idea that wild lands were also beautiful was no longer a foreign concept, and as the top predators who lived there were exterminated—the bears, cougars, and wolves that were more than a distraction for those who traveled in wilderness—the same appreciation for nature's beauty that had developed earlier in Europe began to emerge in North America as well.

It is worth emphasizing that early interest in preserving the natural world was not justified by curiosity about how ecosystems worked or by their importance to human well-being. Ecology as a science was yet to be invented and conservation biology as a distinct scientific discipline would not be formally inaugurated until 1987, when the first issue of the journal *Conservation Biology* was published (Van Dyke 2008). Instead, people began to take note of nature because it was disappearing. In both Europe and North America, people began to recognize that there were limits to untrammeled wildness and those limits were rapidly being reached. It is not surprising that British artists were the first to turn their attention to nature's beauty, because the British were among the first to eliminate wild places from their country. Even in North America, where wilderness was once thought to be limitless, encroaching wagon trains, railroads, large-scale immigration, and burgeoning population growth were well on their way to contracting American wilderness to small remnants of its former vastness. Strategic slaughter was reducing bison herds sixty million strong to biological curios of a few dozen individuals. Market hunting was extirpating passenger pigeon flocks that once numbered in the billions and reducing Eskimo curlews from tens of millions to near extinction.

In fact, the human impact on the land was so complete that entire biomes were functionally eliminated: 950 million acres of eastern virgin forests, including the enormous white pine stands of Wisconsin and Minnesota; the southern longleaf pines of the Southeast; and the deciduous hardwood forests of the Appalachian Mountains were clearcut for lumber and fuel to make way for the agrarian society that had so recently colonized North America. About 99 percent of the seemingly limitless tallgrass prairies of the Midwest was plowed and planted, while the shortgrass prairies in areas of the West that were too dry

Bison were one of many species to be functionally eliminated from North American ecosystems.

to farm were overgrazed by cattle. The natural world was disappearing within a single generation, and people began to grow nostalgic for it. Because they had been forced to wrest a living from wild America, many Americans were proud of their successful struggle against nature. Once their foe, nature had become an emblem of American ingenuity, perseverance, toughness, and superiority. To lose wilderness was to lose an important part of the American identity.

WHERE DID WE GO WRONG?

We reached our low point in environmental stewardship not because humans are evil beings that just like to kill things, but because we never abandoned the adversarial relationship with the natural world that enabled hunter-gatherer societies to survive. The once popular concept of the noble savage, the early humans who lived in harmony with nature, has slowly given way (although not without controversy) to overwhelming evidence that, in developing the ability to run long distances and to throw objects with deadly force, humans evolved to become the most effective hunters the planet has ever known (Liebenberg 2008). It was humans who hunted more than 1000 species of endemic birds to extinction in the South Pacific as they moved from island to island (Duncan et al. 2013). It was humans who eliminated nearly all of the world's large Pleistocene mammals as they colonized each continent (Werdelin 2013; Sandom et al. 2014). This was not an organized goal of human societies, but the result of steady hunting pressure in combination with extraordinarily slow rates of reproduction that characterized these large mammals. And it was humans who began the deforestation of our planet as soon as we learned how to make the metal tools required to do so, and this continues today.

If you think about it, this seemingly ruthless approach to other species made sense. In the old days, it was "nature" that ate us, froze us, starved us, flooded us, killed our livestock, and threatened our crops. The more we beat back nature, tamed it, ate it, or otherwise eliminated it, the better off we were. It was not only ethical to control nature in our landscapes, but necessary, particularly after we transitioned from hunter-gatherers to farmers (Cassils 2004). A "them versus us" philosophy has been deeply embedded in our genetic blueprint from our earliest days, because only those who pushed hard against nature survived to pass on their genes. It was nature or humans, but never both; there was no sharing of space or resources with non-domesticated species—no coexistence at all—and it seemed to work well for humans, until now.

Our lack of reverence toward the ecological networks that supported humans clearly did not acknowledge our ultimate dependence on the natural capital produced by diverse ecosystems, but this rarely mattered, because the earth was huge and humans were few. (See Jared Diamond's 2011 book, *Collapse: How Societies Choose to Fail or Succeed*, for detailed discussions of the cases when it did, in fact, matter very much.) There were usually enough natural resources to supply what we needed, even if we had to move to a new territory to find it. The "slash-and-burn" strategy (still employed by small tribes in the South American Amazon) worked because the earth's forest and grasslands were so huge that overexploited areas easily regenerated before they were burned or overhunted again.

Today, however, there are at least 1600 times more people on the planet than there were before we harnessed agriculture some 10,000 years ago. Not only are there more of us to destroy nature today, but we can now do it faster and more completely. Technological advances have enabled us to eliminate entire biomes at will. We literally move mountains to take what we think we need from the earth, and every day we take more and more, as if the earth were growing along with our own numbers.

THE NATIONAL PARK SYSTEM

Western culture's reaction to the rapid loss of wilderness determined the course of conservation that continues throughout the world today; more countries than ever are committed to preserving the wild places that remain within their borders. Motivated by philosophers like Henry David Thoreau, poets like Ralph Waldo Emerson, activists like John Muir, and politicians like Theodore Roosevelt and Gifford Pinchot, the 1872 U.S. Congress created what would become a global model for conservation: the national park system. The park system was inaugurated with the creation of Yellowstone National Park, widely held to be the first national park in the world. Initially there was some support for managing the natural resources contained within the national park for human use, but the feeling that nature was sacred and could be preserved only in the absence of permanent human settlements prevailed.

In the early 1900s, to encourage visitation to Yellowstone by Easterners (and thus realize considerable economic gain), the Union Pacific Railroad commissioned several paintings celebrating the natural splendor of Yellowstone. The marketing campaign was enormously successful, and soon there was widespread support throughout the nation for preserving wilderness simply because it was magnificently beautiful. The creation of Yosemite and the national park system reinforced this sentiment and helped nurture the climate that enabled Roosevelt to accomplish the greatest expansion of protected lands in history (Brinkley 2009). Citing the justification still heard frequently today, the Roosevelt administration preserved the Grand Canyon and four additional national parks, fifty-one federal bird preserves, four national game preserves, and one hundred fifty national forests—230 million acres in all—to ensure that future generations could experience the beauty, bounty, and wonder of wild places.

Despite the creation of the national park system, national forests, wilderness areas, the U.S. Forest Service, the Bureau of Land Management, and the U.S. Fish and Wildlife Service, and despite the many activities of environmental

Kirtland's warblers are now so few in number that their future is precarious.

The Karner blue butterfly is an endangered species restricted to a tiny portion of its former range.

groups such as the Sierra Club, The Nature Conservancy, the National Wildlife Federation, the National Audubon Society, and land conservancies nationwide, species continued to dwindle and often disappeared altogether from local ecosystems. Populations of birds, plants, and animals—such as Kirtland's and golden-cheeked warblers, Florida panther, marbled salamander, spruce grouse, Winkler's blanketflower, Karner blue butterfly, lynx, bobolink, lake sturgeon, whooping crane, California condor, regal fritillary, rusty blackbird, gopher tortoise, roseate spoonbill, American ginseng, indigo snake, whip-poor-will, Catesby's lily, Florida scrub jay, dozens of fish and mussel species, and many, many more—became too few in number to perform their vital ecological roles or to withstand normal environmental challenges. Other species became critically endangered or disappeared entirely—ivory-billed woodpecker, Bachman's warbler, Columbia Basin pygmy rabbit, dusky seaside sparrow, and others. It became obvious that protecting isolated patches of habitat from development, even if those areas seemed large, was not enough to save many of our nation's plants and animals from local or global extinction. Conservation was not working and we did not understand why.

THE ENDANGERED SPECIES ACT

Although the populations of many species had been slowly declining for some time, the sudden disappearance of peregrine falcons, ospreys, and our nation's emblem, the bald eagle, from areas in which they had once been common finally got the nation's attention. In 1962, biologist and author Rachel Carson exposed these and the losses of other species to the American public in her book *Silent Spring*. After amassing a great deal of evidence, Carson presented a convincing case that chlorinated hydrocarbon insecticides such as DDT were thinning the egg shells of many bird species to the point at which they would break under the weight of a brooding parent. Although *Silent Spring* explicitly focused on impacts to wildlife from pesticides, perhaps its greatest contribution was in recognizing that the things we do where we live, work, and farm—that is, in locales *outside* of protected areas—are critical to conservation efforts everywhere. Because of Carson's writings, environmentalists reasoned that if species were declining despite a large network of parks and preserves, perhaps conservation efforts should be refocused; rather than preserving only the vast wild areas, we should concentrate on saving individual species. This idea became so popular in both scientific and public circles that in 1966, Congress passed the Endangered Species Act.

At first, the act seemed like a reasonable response to an impending extinction crisis, but in practice it has not proven to be the conservation silver bullet many had hoped for. In fact, after nearly fifty years and hundreds of millions of dollars spent on conservation efforts, the Endangered Species Act has become a much-maligned piece of legislation that has been far more successful at turning property owners against the conservation movement than it has been in saving troubled species.

The golden-cheeked warbler, a beautiful bird that breeds only in a small area of oak-juniper scrub in Central Texas, provides a chilling example of the dark side of human nature, as well as the ineffectiveness of the Endangered Species Act. Golden-cheeked warblers were never abundant because of their restricted breeding range, but increasing development in Central Texas in the 1980s triggered a formal review of the warbler for protection under the Endangered Species Act. The reaction of many private property owners was immediate and unfortunate: they cut down the oak-juniper scrub required by the warbler. In fact, to ensure that they would not be restricted in any way by the presence

The golden-cheeked warbler would benefit enormously if property owners were encouraged to provide and maintain oak-juniper habitat, perhaps by offering tax credit incentives.

of an endangered species on their land, many property owners purposefully removed vital warbler habitat from their property, seriously reducing the bird's total breeding range (Dunn and Garrett 1997).

Fortunately, the golden-cheeked warbler has hung on within its diminished range, but the bird's long-term outlook is grim without a dramatic new approach to its conservation. To be fair, the Endangered Species Act has produced some successes when a species' decline could be attributed to a single, easily addressed cause; it did, after all, save the peregrine falcon, osprey, and bald eagle. But the legislation was written in a way that depends exclusively on sticks rather than carrots to protect species, an approach that builds more resistance than support and ignores basic motivational psychology: people respond far more favorably to reward than to punishment. Moreover, the Endangered Species Act requires that we wait until a species has declined to perilously low numbers before any action can be taken. Because tiny populations are highly vulnerable to extinction through normal stochastic (random) fluctuations and inbreeding (Lande 1988; Kearns et al. 1998), this approach almost ensures failure, despite heroic and expensive efforts.

The underlying assumption of the Endangered Species Act has been that conservation and human interests are diametrically opposed, an assumption I will challenge repeatedly throughout this book. After all, as biological entities ourselves, we humans require most of the same ecosystem processes required by other organisms. This brings us to what may be the biggest flaw of the Endangered Species Act: it focuses on saving single species rather than saving the ecosystems on which those species depend. As you will learn, conservation cannot be done piecemeal, because habitat fragments are rarely large enough to accommodate entire ecosystems. This is not to say that conserving habitat wherever we can is not a worthy endeavor; these remnants of the natural world are now the only homes for many species and will serve as essential sources of colonizing species in the future. However, preserving isolated habitats that house endangered species but are surrounded by uninhabitable human-dominated landscapes only postpones the ultimate demise of these species and is a particularly unsuccessful strategy for species that undergo long-distance migrations (Wilcove 2008). Because no species operates independently of thriving ecosystems and the other species within those ecosystems, the potential effectiveness of the Endangered Species Act has been seriously compromised from its inception.

CORRECTING PAST MISTAKES

Our national park system, state parks and preserves, national forests and wilderness areas, and aggressive legislation in the form of the Endangered Species Act have all failed to stem the loss of species, and therefore their contributions, from our local ecosystems. Does this mean that conservation in a human-dominated world is impossible? Was it a mistake to create parks and preserves? Is it impossible to craft legislation that will encourage, rather than discourage, conservation? No, no, a thousand times no! I am convinced that we can practice successful conservation if we think carefully about why our earlier efforts have not worked as well as we had hoped and focus on how to correct past mistakes.

One thing that should be obvious is that we will not succeed if we confine all of our conservation efforts to patches of protected areas. Parks and preserves are central to any large-scale conservation effort, but they will never be enough, because they are not large enough and they are not connected to one another. We will explore the ecological principles behind this fact a bit later in the book; for now, suffice it to say that we need to practice conservation in areas outside our parks as well as inside them. And this means that we need to practice conservation where we live, where we work, and where we farm, because we humans now occupy or have seriously altered nearly all of the spaces outside of our parks and preserves.

We have taken two primary missteps in these early years of biological conservation. The first and most serious has been to assume that people and biodiversity cannot coexist. If this were true, the future of conservation would be bleak indeed, for we humans are rapidly completing our occupation of all of the earth's habitable land masses. By restricting conservation efforts to areas relatively unaffected by humans, we have condemned them to ultimate failure, because such areas are small in relation to the area required for successful conservation and they are isolated from one another. The ecological degradation that defines small isolated habitats is wholly incompatible with successful conservation (Rosenzweig 2003).

Our second misstep has been to leave conservation to the conservationists, that tiny community of trained ecologists who have specialized in the sciences of conservation, restoration, reclamation, and ecological sustainability. It's not that we don't need these people and the knowledge they generate. We do, and we need many more of them. Yet every person on Earth depends entirely on the quality of the earth's ecosystems for his or her continued existence; therefore, each of us, not just a few scientists, carries an inherent responsibility for good earth stewardship. When we leave the responsibility of earth stewardship to a few experts (none of whom hold political office), the rest of us remain largely or entirely uninformed about what earth stewardship is, why we really need it, and how to practice it. Thus, few of us practice any form of stewardship toward the natural systems that support us. I cannot think of a more certain and rapid path to ecological collapse. The conservation of Earth's resources, including its living biological systems (which, unfortunately, are often not considered resources at all), must become part of the everyday culture of us all, worldwide—and it must happen quickly.

CHAPTER THREE

The Importance of Connectivity

Islands are where species go to die.

—DAVID QUAMMEN

ON 12 APRIL 2014, my wife, Cindy, and I embarked on a driving trip to Texas to see and photograph the spring Neotropical migrants as they made landfall along the Gulf of Mexico after their long flight over the Caribbean from the Yucatan Peninsula. This mission included a side trip to Big Bend National Park, which entailed driving more than 200 miles due west on Route 90 through the desert scrub of South Texas. The speed limit was 75 mph, but if you went 75, you were the slowest one on the road. It was my turn to drive, so I powered on at high speed with blessedly little traffic to distract me. At first, we enjoyed seeing the scissor-tailed flycatchers sitting on roadside fences, but soon the land became too dry for the flycatchers and our attention to the passing scenes faded—until, suddenly, we spotted a butterfly, and another, and then another. They looked like . . . yes! They were monarchs making their return flights from their wintering grounds in the oyamel fir forests of Central Mexico! Members of the smallest population of monarchs ever recorded were coming home to create and, with luck, increase the 2014 population as they flew north in search of milkweed to feed their offspring. I turned excitedly to Cindy so we could share one of nature's most phenomenal events, but just as I brought my eyes back to the road, disaster. I hit and killed a monarch—a female that had flown thousands of miles to Mexico the previous fall, had weathered the coldest winter in recent decades, had successfully attracted a male and mated, and had then flown another 800 miles back to the borders of the United States. This magnificent creature had dodged hundreds of thousands of vehicles during her trip south and then again north, but she had not dodged mine as I sped west on Route 90. I was stunned.

SIZE MATTERS

In times past, the death of a returning monarch would have been unfortunate, regrettable, and even sad, but it would not have had a measurable impact on the future of the species. In the past, there were simply too many of them to be concerned with a single butterfly. In fact, many, many millions of monarchs once made the long trek north. A population so large could easily withstand normal environmental challenges and even abnormal mortality caused by vehicle collisions. At one time, there were enough returning female monarchs to create tens of millions more monarchs each summer, which meant that a robust and stable population could return to Mexico each fall. That was then, but it is certainly not the case now. For want of the resources that define monarch habitat—the milkweeds on which monarchs breed, the fall-blooming asters and goldenrods that sustain their fall migrations, and fir forests large enough to buffer winter storms—monarch populations have declined steadily and quickly over the past decade. The overwintering population in 2013, for example, was estimated at only 3.6 percent of the population in 1976, when Canadian botanist Fred Urquhart finally traced the overwintering monarch population to the Mexican mountains. Incredibly, migrating monarchs that were once present in huge numbers have been reduced by well over 90 percent to a tiny remnant of their former population (Monarch Watch 2013).

Fortunately for the monarch, it is beautiful (and we humans love beauty), and its fantastic migration has made it one the most iconic insects in the world, if insects can be iconic. Popular organizations such as Monarch Watch have mounted a national media campaign to encourage Americans to plant milkweeds and create monarch way-stations to aid their migration. In fact, there are now more than fifty partnering groups across the nation whose sole mission is to save monarchs, and these efforts have at least halted the decline of monarch populations. Still, as I write, the monarch population remains relatively tiny, and therein lies the problem: tiny populations are highly vulnerable to environmental perturbations and local extinctions.

This is true for several reasons. The most important reason is that all populations, whether they are tiny or large, fluctuate: in good times they grow, and in bad times they shrink. Let's say, for example, that a population experiences a period of benign weather that reduces physical stress on its members and also creates an abundance of food. This confluence of favorable factors leads to successful reproduction by more females than usual, so the population grows faster than predators can exploit it. In contrast, if the weather turns unusually cold or dry, food becomes scarce, and natural enemies catch up with their prey, reproductive failure in our hypothetical population becomes the rule rather than the exception. The result is a natural decline in numbers that will continue until conditions improve. These cycles in population size are normal and occur over time to a greater or lesser extent in every species on Earth. When a population is large, it typically can withstand even severe environmental insults, but when a population is reduced to a small fraction of its normal size, environmental stress can send it to oblivion.

And this is perhaps the most serious consequence of carving the natural world into tiny and isolated remnants of its former self; habitat fragmentation

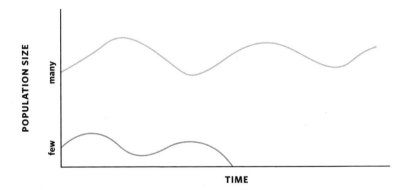

Only large populations (top line) can withstand normal stochastic fluctuations over time, while small populations (bottom line) often decline to zero under adverse conditions.

makes large populations small and isolated from one another, and thus vulnerable to local extinction. When we chop up a large forest into several small woodlots divided by roads, housing developments, strip malls, and other human development, we inadvertently reduce the population sizes of the organisms that inhabited the forest when it was intact. Forest-dependent species with large home ranges, such as the ivory-billed woodpecker, black bear, and elk, disappear from small patches of forest immediately, because each patch is too small to accommodate reproduction by even two individuals, let alone a population of breeding pairs. But even smaller species that can breed successfully in a habitat remnant are put at risk by fragmentation because their population sizes are not large enough to weather the ups and downs of normal population fluctuations, and their forays into the areas surrounding forest fragments, a no man's land for most species, are usually deadly. As a result, inevitable down cycles frequently reduce small populations to zero, and the animals disappear from their habitat fragment forever unless they can recolonize it from another source.

What if we simply made preserves a little bigger? Bigger is certainly better in this case, but populations confined to isolated habitats a fraction of their former size are still at risk from environmental insults, even if such perturbations are rare (Chauvenet and Barnes 2016). And in this era of climate change, extreme weather events are anything but rare. Consider the impact of rainstorms in 2016 and 2017 on endangered Attwater's prairie chickens, which provides a grim example (Elbein 2017). Attwater's prairie chickens once roamed an area exceeding six million acres along coastal grasslands stretching from Louisiana, through Texas, and possibly into northern Mexico. By April 2016, its population had been reduced to 130 individuals confined to two tiny habitat remnants near Houston. Then came the 500-year "tax-day" floods that dropped 17 inches of rain on both preserves simultaneously, destroying all but forty-two wild prairie chickens. The next 500-year flood (or was it a 1000-year flood?) came to Houston only one year later in the form of Hurricane Harvey. This time, the preserves were inundated with more than 50 inches of rain, and breeding populations of wild Attwater's prairie chickens were no more. Doubling or tripling the size of the prairie chicken's preserves would not have been enough to insulate them from such massive flooding. The fate of this subspecies now lies with a captive breeding program that was thankfully ongoing at the time of the storms.

ISOLATION MATTERS

Another species threatened by habitat fragmentation is the eastern box turtle, a species many of us encountered often as kids but rarely see today. Box turtles have a life history that's similar to that of humans; they do not start reproducing until they are well into their teens, and if all goes well, they enjoy a long life—eighty years is not uncommon. Though we usually see them when they are crossing a road or when we stumble across them in a meadow, box turtles actually spend most of their time in woodlots. Males, however, will move between patches of woods in search of females, and females leave the shaded woods when it comes time to lay their eggs. Females bury their eggs several inches below ground in sites where digging is easy and the sun is strong, because they depend on the heat of the sun to incubate the eggs.

In 1968, Paul Catts, a faculty member at the University of Delaware, decided to study the box turtle population in a thirty-five-acre woodlot on university property in Newark, Delaware. The woodlot was, and remains, a typical forest fragment within an urban landscape, bordered on two sides by experimental agricultural fields, on one side by athletic facilities, and on the last side by a four-lane highway. Catts first located and marked all of the turtles in the woodlot—a total of ninety-one turtles. This set the stage for later, as scientists continued to recapture the marked turtles and thus measure how the population of turtles was doing over time.

The study provided a rare chance to follow the plight of a population confined to an isolated habitat fragment. Current land use in surrounding areas has made leaving the woodlot for any reason a deadly activity for the box turtles; they are either mowed or plowed in the ag fields, or they are flattened by traffic on nearby roads. That's why the original ninety-one turtles that lived in the woodlot in 1968 had declined to forty-five turtles by 1994 and just twelve turtles by 2003 (Nazdrowicz et al. 2008). It had taken four decades, but turtle deaths had exceeded turtle births and immigration for so long that the box turtle population was heading to local extinction. Although other woodlots with resident turtles are located within one or two miles of the university woodlot, there is no way the turtles can move between these habitats without being killed. Nothing that could function as a biological corridor, a natural area connecting larger habitats that would enable turtles to move safely between habitats, exists in this corner of Delaware.

Young box turtles spend much of their time underground, like this one, which I accidentally dug up while planting a tree.

If slow-moving turtles were the only species suffering from habitat fragmentation, that would be one thing, but nearly every species that has been studied shows negative effects from carving their habitat into pieces, even when those pieces are quite large (Saunders et al. 1991). Surprisingly, even tiny species suffer. You would think that 35 acres would be enough habitat to sustain carabid ground beetles less than an inch long, but when my student Verryn Jennings compared carabids in the university woodlot with those in White Clay Creek State Park, a 3300-acre preserve only two miles away, he found that twelve of the seventeen species of flightless beetles he collected at White Clay Creek had disappeared from the university woodlot since its isolation (Jennings and Tallamy 2006). It appears that habitat fragmentation is particularly hard on creatures, even the small ones, whose dispersal abilities are limited.

As if small population size and isolation from other viable habitats were not enough, habitat fragmentation introduces other equally deadly challenges to biodiversity. Fragmentation physically changes the habitat patches it creates by increasing edge effects. The edge of a forest is windier, drier, and brighter than its interior. It is also more easily invaded by introduced plants and is far

Today, the greatest threats to box turtles are cars and mowers.

more exposed to predators. Forest-nesting birds, for example, are hammered in small forest fragments by increased exposure to raccoons, crows, cowbirds, and blue jays. When large forests are chopped up into small ones (especially small ones that are long and narrow), the amount of land on the edges of the forest increases dramatically, while the area of the forest interior decreases. In fact, because edge effects can penetrate forests up to 300 feet, many forest fragments have no remaining interior space at all. Quite simply, a forest fragment is no more like a forest than a square cut from a Persian carpet is like a functional rug (Quammen 1996).

Small habitat patches are no place to be when times are hard, but they may have enough resources to meet the needs of at least a few individuals when times are good. Time and again, however, studies have shown that habitat fragments that seem large enough to provide food and shelter for a particular species are empty. This has puzzled ecologists for years: it would seem that small species should be able to use small patches when large species cannot, but it is not uncommon to find habitat fragments that seem large enough to accommodate the breeding territories of several individuals of small, highly mobile species with no one home.

Research on hooded warblers suggests one reason this may be so (Stutchbury et al. 1994; Stutchbury 2007). Hooded warblers, and indeed most vertebrates, have highly specific, sometimes unimaginably complicated requirements for mate choice. Hooded warblers form pair bonds while rearing their young, and for years this species was believed to be monogamous. But when Stutchbury and her students followed both males and females that had paired up, they found frequent dalliances with members of other pair bonds, particularly before eggs were laid. Such extra-pair copulations are so important to hooded warblers that a pair will not set up a breeding territory, even in excellent habitat, if there are not several other hooded warbler pairs nearby. Apparently having the opportunity to fool around on the side is an even more important habitat criterion to these warblers than is high quality food and shelter. In this species—and probably in many more that have not been studied—social constraints, rather than resource constraints, determine the minimal population size and thus minimum habitat size required for successful reproduction.

BUILDING CONNECTIONS

We now live in world in which habitat—places that provide both food and shelter for plants and animals—is so fragmented, and those fragments are so isolated from one another, that they are hemorrhaging species at an alarming rate. I wrote about this in *Bringing Nature Home* (2007) and again in *The Living Landscape* (Darke and Tallamy 2014), but it is such an important phenomenon, and so central to why we need to sustain biodiversity outside of our parks as well as inside, that I will revisit it here.

It is natural for us to view the world from the perspective of what we can normally see around us. When we drive along a highway lined with trees, we subconsciously envision those trees extending for miles beyond the road. That also goes for the woodlot down the street. We cannot see its actual size, but we envision it to be much larger than it really is. All seems well with so much nature out there. What we cannot see, though, is that the strip of nature along the road is an illusion that's only several yards wide. Its function is not to preserve local ecosystems but to block our view of the housing development, industrial complex, or clearcut beyond. The woodlot may be a bit bigger, but it is merely a plot of land with no access that could not be conveniently built upon. Now, however, with the advent of air travel, satellite imagery accessible on Google Earth, and, most recently, drones, anyone who is interested can look at large expanses of our landscape, and even at the entire Earth, in one glance and easily see the actual extent of the human footprint on the biosphere. No more illusions; the shocking human impact is right there for us to see.

Not long ago, my son called to ask me what he should do about the red fox that had built a den under his back porch and was playing with her cute little kits in the backyard. I was delighted and more than a little surprised. My son and his family live in a crowded suburb of Washington, D.C., and I didn't think there would be enough rabbits, voles, and groundhogs in his neighborhood to support a family of foxes. I replied, "That's great! Celebrate! The kids will never forget watching the kits grow up." But he did not share my enthusiasm. "We can't have these things living under the house. They will eat Jimmy" (his two-year-old). I explained that foxes are harmless and that he was very lucky to have such great natural entertainment right in his yard. He persisted. "Dad, this is the real world." I said, "This is a dead world unless we are willing to share it with other living things."

Unfortunately, we have grown so detached from the natural world that my son's reaction to his fox guests is shared by a great many homeowners, urban foresters, and city park managers. "Why do *we* have to save nature *here*? Nature belongs in natural areas, not where people are." This is a good question with roots that have been embedded in our culture for generations, and the answer is not intuitively obvious.

As I have tried to illustrate, we need to restore nature to our home landscapes, to our corporate landscapes, to our municipal parks, and to as much of our infrastructure as we can, because our parks and preserves are not large enough to do the job alone. If our parks and preserves are too small to sustain the species within them, we must make them bigger. But in so many places, all of the land outside of the spaces already protected has been taken for one human enterprise or another. That leaves us three options. We can continue to squeeze the natural world into smaller and smaller places until its inhabitants and the ecosystems they run collapse. We humans could all disappear and let the land we have taken revert back to a natural state. Or we could learn how to share the land we use with other species. My guess is that I will find the most support for option three.

The amount of habitat we have developed for the human enterprise (agriculture, plus cities, suburbs, exurbs, and infrastructure) is so large that you might think that very little quality habitat not already protected could still exist in the lower forty-eight United States. According to the World Bank (2018), a full 44.6 percent of this land is devoted to production agriculture, and this figure soars when the area included in rangeland and tree farms is added. Surprisingly, however, there are still many unprotected pockets of small but high quality land, mostly in rural areas of states east of the Great Plains—Missouri, Arkansas, North Carolina, Virginia, Georgia, the New England states, West Virginia, Pennsylvania—that are privately owned and could play a critically important role in building ecological connectivity if we keep them intact. Fortunately, land trusts and The Nature Conservancy are working feverishly to do just that!

If we connect tiny isolated natural areas by building biological corridors between them, the species that live there can intermingle and increase their populations. And if their populations are no longer tiny, they will no longer be vulnerable to local extinction through inevitable environmental fluctuations. The term "biological corridor" implies that the primary role of these connections is to provide an avenue for safe travel between habitat fragments. But our planted landscapes must become more than areas through which animals and plants can move; they must also be places in which animals and plants can successfully reproduce. In other words, the built landscapes between habitat fragments must be ecologically enriched to the point where they can sustainably support entire lifecycles of local biodiversity. Restoring viable habitat within the human-dominated landscapes that separate habitat fragments—with as much of this land as possible—is the single most effective thing we can do to stop the steady drain of species from our local ecosystems. In fact, restoring the ecological integrity of these landscapes will not only stop species loss, but it will reverse it, and each year we will be able to enjoy stronger, more stable, and more productive ecosystems, as well as more new residents right in our neighborhoods!

Where should we build corridors?

If we are going to turn our planted landscapes into effective biological corridors, we will need to add millions of plants to our neighborhoods, corporate land-scapes, and the lands bordering infrastructure across the country. This needs to happen and can happen even in our densest cities. Where, pray tell, will we find the space to do this? The vast majority of land in the lower forty-eight states is privately owned (U.S. Bureau of the Census 1991), including 85.6 percent of the land east of the Mississippi River, 95 percent of Texas, 94 percent of Maine, and 83 percent of the total conterminous United States. This should make it abundantly clear that many of our corridors will be on private property.

Which parts of our properties are most conducive to building biological corridors? When pondering this question, I eye the landscape now covered in lawns. And I never have to look far to see it. Lawn dominates landscapes in all but our driest ecoregions. Turfgrass has replaced diverse native plant com-munities across the country in more than 40 million acres, an area the size of New England (from figures in Milesi et al. 2005), and we are adding 500 square miles of lawn to the United States each year (Kolbert 2008). A 2005 turfgrass study showed that the small state of Maryland alone had 1.1 million acres of lawn (USDA 2006)—more than twice the area allocated to its state parks, state forests, and wildlife management areas combined (Schueler 2010). Unfortu-nately, these figures are typical of most American landscapes. A few years ago, my students and I conducted an inventory of landscape plantings in twenty-two suburban developments built since 1992 (sixty-six properties in all, randomly selected from within these developments) in southeast Pennsylvania, northeast Maryland, and northern Delaware. Out of the area that could be landscaped (driveways, sidewalks, house footprints, and other hardscapes excluded), an astounding 92 percent was planted in turfgrass!

Much has been written about the precious resources we have polluted or diverted in our quest to grow perfect lawns (Bormann et al. 2001). In the United States, lawn irrigation consumes on average more than eight billion gallons of water daily. In fact, lawn watering accounts for 30 percent of all water used during the summer in the East and up to 60 percent in the West. That's thirty-two gallons of water every day for every man, woman, and child in the country! Because this is more water than is replaced by rainfall in most areas of the country, watering our lawns is clearly an unsustainable practice. What's more, maintaining our lawns in their prestigious, weed-free states has become quite a toxic undertaking (Wargo et al. 2003). All this matters: 40 percent of the chemicals used by the lawn-care industry are banned in other countries because they are carcinogens. Scientists are not guessing about this: Seventy-five studies have documented the connection between lawn pesticides and lymphoma, for example. These same studies show that pets and children are most at risk of contracting cancer, because they spend a lot of time rolling around in the grass. Add to that the connection between fertilizer use and pollution. Homeowners put roughly the same amount of fertilizer on their lawns as is used in agriculture (Law et al. 2004). According to the Environment Protection Agency (2008), 40 to 60 percent of fertilizer applied to lawns ends up in surface and groundwater, where it kills aquatic organisms and contaminates drinking water. If these statistics do not horrify you, then consider the human energy we allocate to lawn care. According to the EPA, we Americans spend more than three billion collective hours per year maintaining our lawns.

But what of the ecological costs? In terms of biological activity, a lawn is the least productive of our plantings, yet it is the default landscaping practice in most spaces. When I think of all that space planted in turfgrass, I cannot help but visualize the forest, savannah, or meadow that used to grow there, as well as the countless bits of nature that once contributed to rich and productive ecosystems. Not only did the replacement of that ecosystem with lawn reduce the ability of the space to support the birds and the bees, but it also reduced the contribution the area was making toward supporting human life. The amount of oxygen produced on a lawn is a tiny fraction of what was released into the atmosphere by the original plant community, and the same goes for the amount of water

Landscapes almost entirely devoted to lawn are all too common in the United States.

cleansed and returned to the underground aquifer, the amount of atmospheric carbon sequestered and pumped into the soil, the amount of topsoil pulverized from the bedrock by plant roots, and the amount of moisture transpired by plants into the air as part of our vital water cycle. The habitat required by dozens of species of native bees for nesting sites was all but eliminated when nature was replaced by lawn, as were the contributions those bees would have made toward pollinating crops and other plants around us. And the myriad species of natural enemies—the predators, parasites, and diseases that normally keep any one species from dominating an ecosystem—were extirpated from the site as well. In short, compared to meadows, shrubs, or trees, lawns are terrible at delivering the essential ecosystem services we all depend on. For those reasons and more, our lawns are great places to start building biological corridors.

Unfortunately, transitioning from lawn to more heavily planted landscapes will not be easy; lawn is one of the most firmly rooted status symbols in Western cultures and has signaled wealth and good citizenship for centuries. Can we really change the lawnscape paradigm?

CHAPTER FOUR

Shrinking the Lawn

[Land ownership is] not just about privilege. It's about responsibility.

—ROY DENNIS

AT A RECENT MEETING of the Pennsylvania Native Plant Society, I met a woman who shared an increasingly common horror story with me. With hopes of bring-ing some wildlife back to her property, she had installed a native meadow in her front yard. I know none of the particulars—its size or design, the location of her house relative to the development in which she lived, her choice of plants, or how well she maintained her meadow—but I do know that something went wrong and her neighbors objected to the finished landscape. One neighbor told her that she had lowered his property value by $10,000! In fact, she was harassed so vehemently that she sold her house and escaped to a more rural setting, where she had no immediate neighbors and she could landscape as she saw fit. In her former neighbors' eyes, this woman had violated the gold standard in suburban landscaping: the immaculate, sterile lawn. She had triggered their belief that anything other than lawn is a low-class product of neglect rather than an artfully designed garden or an attempt to share the land with other living things. Her neighbors did not recognize a natural meadow as a conscious landscaping choice but saw it instead as the abandonment of neighborhood standards. Their status symbol, the weedless lawn, was threatened.

CULTURAL CONSTRAINTS

This woman's unhappy experience with her neighbors takes us to the heart of our proximate motivation for designing landscapes the way we do. Because our gardens are usually in full public view, they are a form of communication. They tell our neighbors whether or not we share their values; and with few exceptions, we work hard to let them know that we do. We conform to the norms of our neighborhood in almost all aspects of our daily lives so that we feel, and are, accepted by our local society. We choose our wardrobes, our cars, our diets, our vacations, and even the plants in our yards with one largely subconscious goal in mind: to send the clear message that we are like those around us and therefore should be accepted as a member of their tribe. Every time we violate the unspoken rules established by the tribe, we risk triggering the suspicion of those around us. Why aren't those people acting like us? If their yard is not just like mine, maybe they don't think just like me? Maybe they don't share my political views or—even worse—maybe they don't believe in my God? Maybe they are one of "Them" instead of one of "Us."

Tribalism

But why do we care so completely whether we are card-carrying members of this all-powerful group? Quite simply, through most of human evolutionary history, being a respected member of a group of supportive people—an extended family, a society, a tribe, a team, a nation—was essential for our survival. Life was almost always tough, and it was much easier to make it through each day safe and sound and with a full belly if we had the support of those around us.

Humans, like many other animals, have always been a territorial species. To survive and reproduce successfully, we needed resources, such as game, water, land, timber, and more, that were often in short supply. It was easier to secure a steady supply of resources if we staked out a territory and excluded humans outside our tribe who competed for the same resources from our territory. (If you doubt your predisposition toward territorial behavior, consider whether you sit in the same seat at the dinner table day in and day out, or whether you sat in the same seat each time you took a class in high school or college. Enough said.) To defend our territories successfully, humans had to be able to recognize those who were part of their tribe and had rightful access to their resources from those who came from outside of the tribe. Someone who looked different, behaved differently, spoke a different language, worshiped differently, or had a different skin color was immediately recognized as an intruder and driven off or killed. We could not tolerate intruders because there were rarely enough resources to feed and clothe our own people, let alone provide for all of those others out there. Being different was a crime, and it was not tolerated.

We are a bit more tolerant today, but as we all well know, the tendency to discriminate against anyone who seems a little different is still very much with us. Does tribalism prevent humans from ever changing preferences? Obviously not; hemlines, hairstyles, and the types of car we drive change frequently and practically overnight at the will of advertisers and popular opinion. Even critically important cultural attitudes that affect foreign policy, our stewardship of the environment, and our view of science can change dramatically and sometimes abruptly. Something in addition to tribal territoriality must influence what we find acceptable and what we do not. That something is our quest for status—status within our family, our neighborhood, our workplace, our bridge club, and our softball team. We seek the respect and admiration of those around us, particularly when we know (or think we know) those people personally. Our madness over driving sport utility vehicles has little to do with our need to drive off-road safely. We do not buy a humvee to avoid flipping over when the next mortar shell explodes nearby. And we do not buy a McMansion because our family of four needs more space. We do these things to send a single message to those around us: We are important people who have made it in this world through hard work and intelligence, and you should admire our high status.

From our earliest beginnings, significant reproductive and survival benefits, what ecologists call fitness advantages, have been associated with having high status within a group. Those advantages are equally real and important today. The marketing industry recognized the power of status long ago. Without threatening our tribal membership, marketing agencies regularly convince us that we can have higher status within the group if we buy this product, watch this movie, use this hair color, wear this suit, drive this car, vote for this candidate, and support these issues.

CHANGING THE CULTURE

People with high status clearly have power over the behaviors of people who want to have high status. Sports and music heroes provide great examples. We afford these successful people incredibly high status, not because we believe baseball, football, tennis, basketball, and rock and roll are incredibly important, but because these people are winners. When basketball star Michael Jordan shaved his head in the 1980s, baldness became cool. But he could not have set this trend in the 1960s, because The Beatles had already told us that long hair would bring high status. When Serena Williams wears a new outfit on the tennis court, Amazon cannot keep up with orders for similar fashions.

Other public figures, such as politicians, command our attention as well, and they always have. For example, when Thomas Jefferson and other respected Americans in the 1700s copied the landscaping paradigms of rich Europeans, creating expansive lawns accented with gardens of exotic plants from exotic places, the standard was set that would dominate how we status-seekers would landscape to this day. Even then, well-kept lawns flaunted two things: The owner was so wealthy he could waste acreage on frivolous turf rather than use it for growing crops or grazing livestock to feed his family. It also boasted of the slave labor or numbers of sheep required to keep the lawn in order (lawn-mowers were not invented until the turn of the twentieth century).

Jefferson could have landscaped effectively and beautifully using native plants, but including native plants would not have achieved his goals. Native plants were everywhere, and using common plants in the garden did not indicate status—they did not require wealth to buy and international connections to obtain. Jefferson featured plants from China and Europe whenever he could, because his landscape was a symbol of his status. Ironically, after more than 300 years, most home gardeners still use non-native plants in their gardens, and this has occurred for so long that we have become masters of mass producing and distributing them. Our attempts to landscape like the rich and famous—to use rare and exotic plants as advertisements of our sophisticated taste and impor-tance—has resulted in cookie-cutter landscapes that are nearly identical. Many horticulturists still scramble to introduce the next exotic status symbol, but in truth they have been far more successful at introducing invasive plant species, devastating insect pests, and deadly plant diseases to which our native plants

Large lawns dotted with spreading trees may remind us of the safe landscapes of the African savannah.

have no resistance. In fact, about 85 percent of invasive woody plant species in the United States are escapees from our gardens (Kaufman and Kaufman 2007).

If our proximate motivation for landscaping with big lawns and spotty specimen trees is to elevate our status within our communities, then what is our ultimate motivation for preferring such landscapes? Why don't we seek status and approval by creating or preserving heavily forested landscapes instead? Why, in other words, was the standard for acceptable landscaping already established by Europeans when Jefferson pondered the landscape design of Monticello? No one knows for sure, but evolutionary psychologists have offered an intriguing hypothesis that explains why cultures the world over transform natural landscapes into open, savannah-like settings (Cronk 1999). Perhaps modern humans prefer landscapes dominated by large expanses of short grass because ancient humans felt safest in such settings.

Our lineage began in the vast savannahs of Africa, where humans all too often became dinner for lions, leopards, cheetahs, hyenas, wild dogs, and other proficient predators that were common at the time. Early humans' best defense was to avoid these animals by being able to spot them first: the shorter the grass and the fewer the trees and shrubs, the better the chances of making it through the day. Humans chose habitats with a few large, spreading trees with low and accessible limbs, because being quickly able to climb such trees provided a means of escape in case of an animal attack. Evolutionary psychologists posit that today we still feel more at ease in landscapes that afford unobstructed views of our surroundings, and because of this, we have transformed forested ecosystems throughout the globe into managed replicas of the African savannah. A few anthropologists challenge this hypothesis by pointing out that not all cultures value open landscapes, such as the Yasuni people of Ecuadorian rainforests, and therefore preference for such spaces must be culturally determined rather than inherited from our ancestors. (Others counter that the Yasuni are descendants of refugees who escaped Spanish slavery by choosing the dense forests of the Amazon as a hiding place [Mann 2012].) If such cultural preference is true, this would be good news, because culture is surely easier to change than genetic predisposition. In either case, however, whether sprawling landscapes provide us inherited comfort or are simply an artifact of cultural values, they have come with a huge price tag: most of the plants and animals with which we share this planet cannot survive for long in savannah-like landscapes, and these species are disappearing from human-dominated landscapes by the thousands.

Regardless of whether our motivation is culturally, ecologically, or evolutionarily influenced, is it possible to change such deeply entrenched preferences? Of course! In fact, we've already made many cultural changes to reflect changing times. We changed cigarette smoking from a high-status social activity to an embarrassing and nasty habit banished to parking lots. Activism by numerous celebrities and animal rights groups helped change our cultural view of seal

Grass pathways, such as this one at Mt. Cuba Center in Hockessin, Delaware, can guide pedestrians through a well-planted landscape to help them avoid trampling tender plants.

and other exotic fur coats, from expensive and sought-after luxury items to products banned for sale in the United States and elsewhere. Most American women in the mid-1950s were housewives; today, women make up about half of the total U.S. work force. Such cultural changes don't occur overnight, but they clearly illustrate the power of peer pressure (the collective will of tribe), especially when backed by compelling logic. And when logic-based peer pressure is combined with financial incentives, it is irresistible.

Many of us will do nearly anything if our peers are doing it, especially if it saves us a little money. With regard to landscaping, both of these forces are currently at work in the water-starved American West. Growing a lawn in areas with insufficient rain requires enormous amounts of irrigation. The logical pressure to switch from growing water-thirsty lawns to xeriscaping with plants that require little or no water is overwhelmingly compelling. When municipalities in Southern California and Arizona offered to pay homeowners to remove lawn, for example, the lawn culture suffered what may be a fatal blow in desert biomes. Now, status in those areas is associated with the person who landscapes with the greater good in mind rather than the gardener who wastes limited water supplies trying to grow grass in a desert.

I see similar landscaping changes starting to occur for ecological reasons, even in areas of the country that receive lots of rain. The realization that the formula for successful conservation involves less lawn and more native plant communities is the compelling logic that is motivating our reassessment of plant choice and abundance in our landscapes. We know that our native plants make wonderful garden ornamentals, because, after all, they are now favored by high-status landscapers in Europe. And we know that, when used properly, plants native to our area will do well because they have thrived there without our assistance since the last glaciation. All that remains is to attach the same respect and admiration—that is, the same status—we used to associate with exotic plants to the use of native plants. As the essential relationship between native vegetation and sustainable ecosystems becomes common knowledge in our dawning age of environmental stewardship, and as more and more high-end properties demonstrate how effectively and beautifully natives can be used in attractive designs, landscapes dominated by natives will become the norm instead of the rare exception.

Reducing the area we currently allocate to lawns is also a necessary and logical consequence of ecological landscape design. I don't think it's likely (nor do

I suggest) that we will ever abandon the lawn as a landscaping tool. Turfgrass species are perfect for areas where we walk, for example, because they can withstand moderately heavily foot traffic. But transitioning from landscapes in which wall-to-wall turfgrass is the default, to landscapes that thoughtfully use lawn as pathways through savannahs of spreading native trees, native forbs, and warm-season wild grasses is now entirely within our grasp and presents a new way to demonstrate our creative abilities.

CHAPTER FIVE

Homegrown National Park

There can be no purpose more inspiring than to begin the age of restoration,
reweaving the wondrous diversity of life that still surrounds us.

—E. O. WILSON

SEVERAL THINGS POINT us toward a new approach to conservation.

- Conservation that is confined to parks will not preserve species in
  the long run, because these areas are too small and too separated
  from one another.

- Although we must continue to protect good habitat wherever it still
  exists, we can no longer afford to ignore the ecological value of the
  land outside of our preserves—that is, the areas between isolated
  habitat fragments.

- Restoring habitat where we live and work, and to a lesser extent
  where we farm and graze, will go a long way toward building
  biological corridors that connect preserved habitat fragments with
  one another.

- Creating biological corridors will enlarge the populations of plants and animals within protected habitat, enabling them to weather normal population fluctuations indefinitely.

- Across the United States, millions of acres now covered in lawn can be quickly restored to viable habitat by untrained citizens with minimal expense and without any costly changes to infrastructure.

What if each American landowner made it a goal to convert half of his or her lawn to productive native plant communities? Even moderate success could collectively restore some semblance of ecosystem function to more than twenty million acres of what is now ecological wasteland. How big is twenty million acres? It's bigger than the combined areas of the Everglades, Yellowstone, Yosemite, Grand Teton, Canyonlands, Mount Rainier, North Cascades, Badlands, Olympic, Sequoia, Grand Canyon, Denali, and the Great Smoky Mountains National Parks. If we restore the ecosystem function of these twenty million acres, we can create this country's largest park system. It gives me the shivers just to write about it. Because so much of this park will be created at our homes, I suggest we call it Homegrown National Park.

Homegrown National Park will differ from traditional national parks because it will not be confined to a single location. It will be everywhere, helping to preserve lives and ecosystems in all bioregions and all biomes—from the northern coniferous forests of the Northeast to the desert biomes of the Southwest; the Coastal Plain Pinelands of the Southeast to the Douglas fir forests of the Pacific Northwest. We will not confine our native plantings to our backyards, as so many backyard habitat programs suggest we should; that would eliminate half of our targeted land before we even start. It would also imply that our new park is so ugly we have to hide it from public view in the backyard. Not so! Planting a stately oak tree and other attractive native plants in your front yard will add immeasurably to your local ecosystem, and few would object to their presence.

Creating Homegrown National Park will require a collective effort from landowners everywhere. It will require the largest cooperative conservation project ever conceived or attempted. And it must be voluntary rather than mandated. How will we ever bring so many people onboard? People who realize the conservation potential of residential landscapes, city parks, hedgerows, roadsides, and other unnatural human constructs tend to do so in classic, head-slapping "Ah ha!" fashion. Quite suddenly, the solutions to many of the environmental issues that have grown in proportion to our human footprint become clear. What hinder these solutions are the tens of millions of people who, through no fault of their own, remain clueless. So how do we move beyond preaching to the choir to reach the uninformed masses?

It is ironic that spreading good ideas should pose a challenge in a world in which our ability to communicate is instantaneous, global, cheap, and nearly constant. Why can't we tweet, Instagram, Snapchat, YouTube, and TED Talk our way into the hearts and minds of the millions who do not have enough experience with the natural world to know that it is a wonderful place—and, oh yes, to know that we will not survive its demise? Maybe we can, but it would require more than 280 characters or a ten-minute video to change most adults' value systems. And that truly is the challenge—to convince people that, for their own good, they need to value something they do not currently value.

Consider how miserably we've met the challenge posed by climate change. Thousands of studies by the best scientists on the planet, dozens of feature-length documentaries, decades of international congresses, and global media attention have failed to convince 66 percent of U.S. citizens that a problem even exists, let alone the urgency of enacting solutions (Pappas 2015). It's not that the message is too difficult to understand (we are fouling our own nest), or the evidence too elusive (before and after pictures of glaciers are as unambiguous as you can get). Rather, the simple message that we are changing the planet's climate in very unpleasant ways has been purposefully enmeshed in the tribal identities of political right and left, conservative and liberal, red and blue. Climate issues have been labeled "liberal issues" and therefore are not allowed to be considered in isolation by conservatives without threatening all conservative ideology. We have witnessed the same ideological entrenchment on the left regarding childhood vaccinations and genetically modified agriculture. Emotion quickly takes the reins, and no litany of facts can convince us that our tribal dogma is incorrect.

How, then, can we hope to convince a tribally constrained public that biodiversity is essential and must be saved not only in its current state, but it must be restored to much of its prehuman glory in our yards? One powerful approach might be to provide the means by which the public can discover the rewards of responsible landscaping for themselves. It is our nature to resist ideas that are forced on us, regardless of their value or wisdom, especially when they come as mandates from governmental institutions. But if we discover the truth in these messages ourselves, tribal loyalties no longer cloud our vision and we can become enthusiastic proponents instead of bulldog obstructionists.

HOME SCHOOLING

Here's an example of what a powerful learning tool self-discovery can be. Recently a friend described her father as being a stereotypical proponent of traditional landscaping. To him, the ideal landscape looked just like his neighbor's yard: beautiful, pruned and mowed, always perfect, and inanimate. A large immaculate lawn was a sign of responsible neighborhood stewardship—the sign of a good citizen, a team player who was doing his or her part to maintain local property values. Needless to say, her father was not a member of the Bringing Nature Home choir.

My friend did not share her father's landscaping values and often urged him to think beyond what plants looked like and consider instead what they could do for his yard's watershed, soil, pollinators, food webs, and carbon sequestration. Alas, he would have none of it, so she turned her attention elsewhere. Concerned about the rapid decline of monarch butterflies, she joined Monarch Watch, a national organization that promotes the creation and restoration of monarch habitat and provides the host and nectar plants needed to rebuild monarch butterfly populations. She planted milkweeds and asters in her yard, local parks, and other open spaces, and became so involved that she often asked her father for help minding her young kids. Finally, he asked what was taking so much of her time. She explained that monarch populations had declined by more than 96 percent in ten short years because of the loss of milkweeds and forage plants in both agricultural and residential landscapes, and she was working to reverse those losses. A few days later, she was flabbergasted when she received a call from her father asking whether she would be willing to plant some milkweeds in his yard. Naturally, she jumped at the chance and planted a row of milkweeds along his back fence.

In a few days, her "flabber" was even more "gasted" when she started getting regular calls from her father: "I saw a monarch!" "A monarch laid an egg!"

Monarch caterpillars develop exclusively on milkweeds and their close relatives.

"I don't think there is enough food for the monarch caterpillars. I need more milkweed plants!" And so it went the rest of the summer. He was hooked! Her father had transformed in mere days from someone utterly detached from the natural world to someone enthralled with it, all because he had gained personal experience with a beautiful butterfly that is in trouble. He recognized that, through his actions, he had become part of the solution instead of part of the problem, and he felt good about that. Monarchs had awakened his paternal instincts, and he nurtured those beautiful creatures as if they were his own. He learned that his beloved monarchs could eat only milkweeds; his Asian ornamental species—his crape myrtles and goldenraintrees, barberries, hydrangeas, and Callery pears—provided nothing for them. The grass he once dutifully mowed weekly was an impediment to the milkweeds he needed for his monarchs, so he replaced some lawn with more milkweeds. Perhaps his greatest epiphany was realizing that if he could use his property to help save monarchs, he could also use it to save zebra swallowtails, Carolina chickadees, dozens of native bee species, and countless other creatures, all by adding the plants on which these animals depend. My friend's father now is drawn into his yard not by the competitive need to out-manicure the Joneses, but to learn, to watch, and to be fascinated and entertained. To the devil with the Joneses! His loyalty now lay with the monarchs.

PERSONAL GAINS

This man experienced a new relationship with nature, perhaps for the first time—what so many have experienced when they visit our magnificent national parks—but he did it right at home. It might seem preposterous to equate a walk in your yard with a trip to Yellowstone. Your yard surely will not provide breathtaking views of the Grand Canyon of the Yellowstone, Bridal Falls, or the Teton Range, but there is much that Homegrown National Park can provide without the expense, crowds, reservations, or traffic jams of a monumental road trip. And it can provide these things every day!

Just as we divide our houses with structural walls into discrete rooms, we can create private spaces outdoors with walls of vegetation.

The beautiful zebra swallowtail is one of many creatures that can entertain you during quiet times outdoors.

CLOCKWISE A spring peeper's call signals the arrival of spring; a gray tree frog ushers in the summer season; a white-lined sphinx moth pollinates a flower while seeking nectar; a dark-eyed junco signals the arrival of fall as well as any calendar.

If you are seeking peaceful solitude in a natural setting, a yard with plantings that create outdoor rooms is ideal for escaping a media-crazed world, or even just the scrutiny of your neighbors. A bench surrounded by privacy screens built from a diversity of woody plants—what landscape designer and author Rick Darke refers to as organic architecture—can provide escape from public display, the opportunity to think (or not think), or a chance to return to your roots in the natural world with all of its fascinating components. Position your bench with a view of the hole in a tree, where a male white-breasted nuthatch

 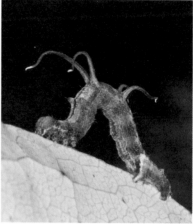

The flamboyant markings and paddle-shaped bristles of the paddle caterpillar challenge the imaginations of kids and adults alike.

The filament bearer geometer sports bizarre tentacle structures on its back, prompting questions and wonder.

is performing his stylized, circular dance, wings outstretched and head arched back, in an attempt to impress a difficult-to-impress female nearby. While you sit, you can marvel over the strength and tenacity of a small black spider wasp as she drags a paralyzed spider twice her size over any and all obstacles to an underground burrow, where she will lay a single egg on it and seal the entrance hole so her offspring can develop out of harm's way. Look closely at the dead leaf on your staghorn sumac; it may very well not be a leaf at all, but the caterpillar of the showy emerald moth. Or simply watch the parade of butterflies, native bees, beetles, wasps, ants, and flies seeking nectar from your blooming sweet pepperbush. And what you see today will be different tomorrow, as the seasonal progression of the species in your yard changes daily. Such experiences with nature can certainly be sought and found within an existing national park, but typically they are not. These experiences take time, convenience, patience, solitude, and serendipity—things usually absent from harried vacations.

Admission to Homegrown National Park is free, and there are no restricted seasons. As you become familiar with the natural cycles that occur in your yard, you will start to anticipate them, subconsciously at first, but then as something you eagerly await. Cindy and I rely more on our yard than on the calendar to tell us when the seasons are changing. Spring does not arrive at our house until the spring peepers, American toads, and woodcocks signal its arrival with their mating calls. Early summer is ushered in by our new resident, a gray tree frog, while the dog days of August bring the white-lined sphinx moth to pollinate our evening primrose. We know it's fall when the dark-eyed juncos and white-throated sparrows take up residence, and it's early winter when the great horned owl hoots away the late-night hours. We look forward to these events, and we welcome each of our friends with a smile when it first appears in our yard. But our real joy comes from the message of hope each brings to us: nature is still happening all around us.

Cultural geographers, anthropologists, sociologists, and especially horticulturists all write about a sense of place—the personal feeling of identity, comfort, and special meaning that certain places hold for us that other places do not. A sense of place may come from early childhood experiences or from long years spent in a particular place where many defining moments in our lives have occurred. I am willing to bet that by helping to build Homegrown National Park, you will build a connection to place more powerfully and quicker than you ever have in the past. By restoring the plant and animal communities that belong where you are, you will develop an intimate connection with each community, like that sense of parental responsibility deeply experienced by the monarch-loving father of my friend. We can never truly own nature, but a sense of ownership creates a strong stewardship ethic, something the land we occupy desperately needs.

THE THRILL OF DISCOVERY

Humans are, by nature, an adventuristic species. We crave new experiences, new horizons, and uncharted territories. Humans are also colonists at heart. The human species arose in Africa, but that continent wasn't big enough for us, so we moved into lands that are now Eurasia and Australia, and then on to the Americas. We have now explored and colonized all of planet Earth, so we are looking upward, as now, rather foolishly I think, we talk of colonizing Mars. Though our great explorations are over, most of us still yearn to discover. We have replaced wagon trains with road trips in hopes of satisfying this age-old craving, but we rarely succeed. One city sprawl looks pretty much like the next, despite efforts to convince us otherwise. But consider this: if it is no longer easy to experience the thrill of discovery by finding new places, perhaps we can

Zoe, the lizard stalker

derive the same exhilaration by discovering new things—new knowledge, new experiences, and new sightings. The popularity of listing new bird sightings certainly attests to the excitement people derive from nature. In fact, the boun-ties of the natural world offer endless opportunities to enjoy new experiences, especially if we bring those bounties to our own yards.

No one is more inherently excited by other living things than our children. To them, nature is a fascinating, totally unexplored new world that challenges their active imaginations at every step. Finding the bizarre paddle caterpillar or the equally odd filament bearer geometer, for example, will inevitably lead to the question, "What are those paddles and filaments for?" Good question, but don't answer it. Let your kids work out some answers for themselves. It's fun and helps them think about the daily challenges other creatures face in our gardens.

My granddaughter Zoe illustrates how creative kids can be when they enter the captivating world of nature. Zoe's yard is miniscule, perhaps 300 square feet in all, but it provides plenty of room for her to demonstrate how to stalk lizards. According to Zoe, you must first disguise yourself with sticks and leaves so the lizard will not see you. Then you creep ever so slowly along the ground toward the lizard. Patience is key; any sudden movement will send those lizards scampering. Never smile or laugh. Your white teeth will definitely scare the lizards away. Why you are stalking the lizards is not important; the fun is in the stalking. And, by the way, this technique works well whether you are wearing your new dress or your jammies.

Frequent visitors to your yard, like this brightly colored cardinal, can become as familiar to you as your pets.

NURTURING OUR NATURAL PETS, AND OURSELVES

When a landscape is transformed from a self-sustaining native plant community to a suburban lawnscape, the cardinal in your yard is not just a cardinal in your yard: it is *your* cardinal. As such wild creatures can no longer depend on wild natural plants to sustain them, you must assume responsibility for the well-being of your cardinal, your blue jay, and your American toad. All these creatures that once lived in a wild landscape now depend on you to meet their needs. Fortunately, it is easier to care for local wildlife than it is to take care of more traditional pets. In fact, you can transfer this responsibility entirely to plants if you rebuild your yard into a Homegrown National Park with well-chosen natives. You can rebuild some or much of the food web that once existed in your yard by including plants that provide nourishment, cover, and forage for local wildlife, and they might not even notice that you live among them. Close interactions with the wild animals in your yard can bring you the same emotional benefits that are gained from living with cats and dogs.

To the surprise of the public, and even to researchers, studies have shown that brief exposure to the natural world produces measurable medical and social benefits for humans (reviews by Kuo 2010; Wolf 2014; Wang and MacMillan 2013; Louv 2012 and 2016; Cracknell et al. 2016; Kardan et al. 2015). Plant a tree outside a classroom window and test scores improve. Plant a tree outside a hospital room window and the patients in the facility heal faster. Studies show that apartment buildings with treed courtyards house families that undergo fewer divorces, higher graduation rates, and less juvenile delinquency than nearby apartment complexes with no trees. Spending just fifteen minutes in a peaceful, natural setting reduces our blood pressure as well as the levels of cortisol, the stress hormone, in our blood. What, pray tell, is going on?

There is a growing consensus that all of these effects stem from the reduction of stress that results from contact with nature—what Richard Louv calls "Vitamin N" (2016). Quite simply, when we experience less stress, we do everything better—from learning, to healing, to interacting with others. We know that the presence of trees, natural plantings, and the butterflies and birds they support reduce our stress levels. At this point, we can only speculate as to why. One key finding that is particularly relevant to Homegrown National Park is this: The extraordinary health and social benefits we derive from exposure to nature are short-lived and realized only from repeated exposures. A two-week visit to Yellowstone National Park during the summer will not reduce our stress levels for the rest of the year. The only way we can benefit from Vitamin N in a meaningful and sufficient way is to live or work within a natural setting, or to visit one on a regular basis.

LEARNING STEWARDSHIP

Louv struck a chord, particularly with baby boomers who grew up playing outside, when he wrote *Last Child in The Woods* in 2008. He noted the many ways generations of young people have benefited from the freedom to discover the world, particularly the natural world, at their own pace when they were ready. Powers of observation, curiosity, wonder, appreciation, independence, that precious sense of place, and, most important, the building blocks of knowledge

A short walk in the woods can restore our attention span, make us more hopeful and compassionate, and improve our mood, while the presence of street trees can reduce the frequency of crime in an area and improve the cardiometabolic conditions reported by residents of those streets.

about how the world works are all byproducts of unsupervised play in the woods, meadow, and pond next door. Today, of course, there are no woods, meadows, or ponds next door for most kids, and if these natural areas are nearby, unsupervised play within them is unheard of. Parents who shoo the kids out the door in the morning with instructions to be home by lunchtime are more likely to be arrested for child abuse than praised for good parenting. (I wish I were kidding.) These days, if life lessons are to be learned, they will be likely learned through apps and websites, and the learning will happen within easy reach of an electrical outlet.

Nature has become irrelevant to our kids and to most of their parents as well. In the United States, 82 percent of us live in cities with artificial environments that, on the surface, seem to be self-sustaining. The unspoken message that is sent to city-dwellers day in and day out is that what happens outside of cities does not affect us and therefore is unimportant. From a conservation perspective, a near total cultural disconnect from nature is a big enough challenge, but we have gone one step further. We have demonized nature. We no longer talk of a stern but nurturing Mother Nature, whom we should hold in reverence and upon whom we are all completely dependent. Instead, we are told by sensationalizing news media that if we go outside we are unlikely to survive—we are sure to be attacked by a mountain lion, maimed by a coyote, struck down by West Nile or Zika virus, disfigured by poison ivy, or crippled by Lyme disease.

To be sure, such risks do exist, because life is not and never will be risk free. But the actual numbers tell us that the dangers lurking in nature are so miniscule compared to the dangers we ourselves have created through technology that it is foolish to avoid a relationship with nature for safety purposes. Our misguided risk assessment has convinced us that tailgating a forty-ton tractor trailer at 75 mph is of no concern, but revegetating our yards will surely result in a deadly bite from a rattlesnake species that in reality was extirpated 150 years ago. We simply do not believe that we are far more likely to die from catching the common cold, falling out of bed, slipping in the shower, tumbling down the stairs, or using a toaster, vending machine, cell phone, or prescription drug. These dangers were not part of our evolutionary past, so, regardless of how real they are today, we ignore them.

There are many consequences of today's societal disconnect with nature, but one stands out from the rest because of its potentially catastrophic consequences: our kids are the future stewards of planet Earth, and they are woefully unprepared for this awesome responsibility. Most young people don't know that the earth now needs constant stewardship—or what that stewardship even looks like. Even fewer know that a hands-off approach to good earth stewardship is not the best option, because there are few places where nature is still able to take its course without human management. Many kids don't know these things because we have not taught them and they have been deprived of the chance to learn about the natural world on their own. Fortunately, Louv's *Last Child in the Woods* and subsequent books have triggered a national—indeed, a global—discussion about how to rectify this alarming situation. How can we get our kids (and ourselves) back to nature when so little nature is still present in our world? How can we teach them that nature is a wonderful place to be and its risks, which are easily managed with a little knowledge and respect, are far outweighed by its benefits? How can we expose young people to the millions of other species on this planet long enough for them to develop a caring relationship with them?

One of the most obvious responses has been to take kids to nature. Organize a field trip to the nearest state park and take them on a hike (with an adult, of course—but make sure those little ones stay on the path and, for heaven's sake, don't touch anything). Show them the wonders of nature, but do it in three hours and then get back on the bus and return to the classroom where real learning takes place. And for logistical reasons, confine the trip to the midday hours—never dawn or dusk when nature is most apparent and at its best. My point is a simple one: although organized exposure to nature is far better than no exposure at all, it is unlikely to achieve the goal of enabling kids to develop a personal relationship with the natural world. School trips, community hikes, and even family picnics are social events involving other people—often many other people. They are typically highly structured and regulated, and they are far too infrequent to help children develop a sense of the seasonal changes or the many natural cycles that often go unnoticed every day. Scheduled expeditions to parks lack the solitude, the frequency, the extemporaneous opportunity, and the unhurried exposure to natural things that can come from a lazy afternoon within the safety of their own yard—if nature is in their yard.

To me, one of the biggest benefits of Homegrown National Park is providing our future earth stewards with the convenient option of entering the natural world 365 days of the year right at home. The simple, undramatic, and commonplace encounters with the plants and animals in our yards will convey the sense of responsibility we all must have toward protecting and nurturing them. In short, Homegrown National Park will teach us, and our children, to value the natural world rather than destroy it.

CHAPTER SIX

Rebuilding Carrying Capacity

In the future, we must be more ecocentric than egocentric.

—SUSAN LERNER

MY WIFE AND I RECENTLY ENJOYED a family visit in Portland, Oregon, a city known for its leadership in energy-efficient building design and eco-friendliness, and one that boasts the largest park in a major city in the United States—Forest Park. Portland is Green, with a capital G, and many would argue that it is our greenest city. Nevertheless, it didn't take long for me to notice that there seemed to be very few indigenous species among Portland's abundant street trees. I saw lots of Asian species such as Callery pears, ginkgos, and zelkovas; European species such as Norway maple, little-leaf linden, and horse chestnut; and even trees from eastern North America such as sweetgum, red oak, and sycamore. Lest I was imagining it, I enlisted my granddaughter Sofia to help me inventory Portland street trees in the Sellwood and Hawthorne neighborhoods where our kids live.

After a few days of wandering the sidewalks, we had identified 1176 trees; as I had suspected, only 100 of them were indigenous to the Pacific Northwest. The rest, a full 91.5 percent of the trees in those neighborhoods, were introduced from other continents or ecoregions. That means nearly all of Portland's

street trees lack the evolutionary history with local wildlife that is required to contribute meaningfully to the food webs that support them. If our small survey represents trees in all of Portland's built landscapes, then the city is a great example of how to create an attractive city with very few breeding birds, butterflies, bees, or other desirable wildlife.

I'm not disparaging Portland. In terms of environmentalism, the city has led the way in many respects, and I have little doubt that Sofia and I would find similar results if we surveyed trees in other cities. My point is that even our greenest cities have missed the most critical aspect of nature-conscious urban design: plant choice matters! Though they pride themselves in their green and sustainable lifestyles, Portlanders' choices of non-native landscape plants have lowered the carrying capacity of the city to the point that, during the bird breeding season, it is a silent city. How can we consider something sustainable if it doesn't sustain life?

That said, some birds have figured out how to make a living in Portland. Crows and scrub jays can breed within the city limits because they can survive on human garbage. Robins can breed there because they can rear their young on the earthworms they find in lawns. But birds that are abundant just out-side of Portland, such as spotted towhees; white-breasted and red-breasted nuthatches; black-capped and chestnut-backed chickadees; downy, hairy, pileated, and Lewis's woodpeckers; western bluebirds; ruby-crowned and golden-crowned kinglets; Oregon juncos; western tanagers; black-headed gros-beaks; black-throated gray warblers, Bewick's wrens, Steller's jays, and others breed only rarely within the city of Portland and its suburbs. My guess is that these birds are eager to breed in Portland, and during the winter when people put up bird feeders, many do come into the city to eat. But most birds rear their young on insects rather than seeds or berries, so when it comes time to reproduce, the plants available in Portland simply do not make enough insect food for birds to raise their young successfully.

LIMITS

Between 2001 and 2005, 1360 scientists from around the world collaborated to assess the state of Earth's ecosystems and their ability to produce the services

A street in Portland lined with European little-leaf linden

CLOCKWISE Steller's jay; Lewis's woodpecker; Chestnut-backed chickadee

that sustain us—the clean, oxygen-rich air; clean, accessible water; flood control; weather moderation; stable fisheries; rich, fertile topsoil; forest products; pollination services; carbon sequestration; natural sources of pest control; and other services. The end product, the Millennium Ecosystem Assessment, was published in 2005, and its message was grim: by the time the twentieth century yielded to the twenty-first century, we had already destroyed 60 percent of the earth's ability to support us. That is tantamount to shrinking our planet by 60 percent while we continue to expand our populations, our economies, and our needs. Equally alarming is that we have not stopped overexploiting the earth; conditions have continued to deteriorate in the ensuing years since this study.

Most of us have trouble connecting our daily lifestyles with the depletion of limited natural resources. After all, most of our basic needs—for water, air, and food, for example—are met when we need them, so they don't seem limited at all. I have been as guilty of this oversight as anyone, especially when I

Mountain gorillas, like every species of sea turtle and thousands of other creatures are declining rapidly because of competition with humans for the earth's finite resources.

was a child. When I was eight years old, I joined the swim team at our local YMCA. I don't remember much about the workouts, except that the water was cold. I do remember the long, hot showers I would take after each practice. I would stand forever under the steaming water, mesmerized by the warm and massaging flow pouring over my head. I was always the last one out of the shower. Thinking back, I must have leveled a mountain of coal and drained an entire reservoir to create enough hot water for my luxurious showers. And there was never any talk of limits—except the limits to my father's patience while he waited twice a week for me to emerge from the locker room. No talk of how taxing my indulgence was on two precious resources, water and energy.

Unfortunately, the disconnect between what we choose to do and the resources with which we use to do it is still very real today. A surprising number of people simply do not accept the notion that we humans have exceeded the earth's ability to sustain us in the long term. Many believe that technology will always provide solutions to resource shortfalls and that the concept of a finite earth is simply an alarmist view. But think about it: if we were not using resources faster than the earth was replenishing them, we would not be hearing about running out of fresh water in South Africa, the Middle East, and our own Southwest. Nor would we be running out of topsoil, copper, rare metals, phosphorus, fish stocks, arable farmland, and dozens of other essential resources. Our rivers would still reach the sea, and we would not have already lost 800 species of plants and animals to extinction. In North America, we would not have 8500 species of plants and animals (more than a third of our best-known species), including 432 species of birds at risk of extinction (North American Bird Conservation Initiative 2016; Tolmé 2017). There would be no need to worry about polar bears, tigers, snow leopards, jaguars, condors, sea turtles, orangutans, mountain gorillas, elephants, rhinos, giraffes, blue-fin tunas, sharks, cod, and countless pollinators. We would not be faced with losing 47 percent of our plants, 40 percent of our fish, 25 percent of our mammals, 33 percent of our amphibians, 35 percent of our reptiles, 13 percent of our birds, and 52 percent of our insects. In short, we would not be looking at the sixth mass extinction in the history of life on Earth, and the first to be caused by a single species (Kolbert 2014). There would be plenty of resources, including the space needed to exist, for all humans, plants, and animals.

CARRYING CAPACITY

Carrying capacity is the ability of a particular place to support a specific species; it refers to the number of individuals of the species that can be supported indefinitely without degrading the local resources. The 2005 Millennium Ecosystem Assessment, in effect, measured the earth's carrying capacity with respect to humans. Long ago, when there were far fewer humans on the planet, the resources required for them to survive and reproduce were replaced by many vibrant ecosystems faster than they were used. In other words, ancient humans' relationships with the ecosystems that supported them were sustainable and could go on, at least in ecological time scales, forever. In contrast, when a local ecosystem is degraded by one or more species, we say those species have exceeded the carrying capacity of that ecosystem and the relationship is

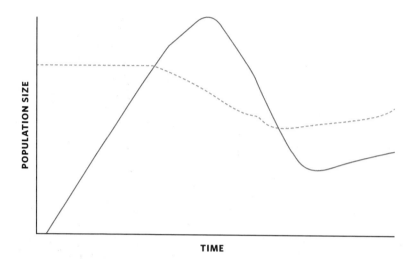

As soon as a population (solid line) exceeds the carrying capacity of an area (dashed line), the resource base is degraded and the carrying capacity declines.

not sustainable. A population that grows larger than its carrying capacity uses resources faster than they are replenished by nature; as a result, the quality of the habitat is diminished (that is, the carrying capacity is reduced) and the population declines until the local environment has repaired itself.

We might wonder, for example, what is the carrying capacity of white-tailed deer in the Eastern Deciduous Forest biome? The answer would be ten to fifteen deer per square mile, depending on the quality of the forest, the amount of edge habitat available, and the availability of the browse that deer require. But if the carrying capacity for deer is so low, how can seventy, eighty, or even one hundred deer per square mile survive for decades in my Pennsylvania county and in so many other places? Even more perplexing, why do all those deer seem healthy?

This is where the second part of carrying capacity's definition becomes important: the number of individuals of a species that can be supported indefinitely, *without degrading the local resources*. There may be far more deer in the East than the estimated carrying capacity says there should be, but they are not existing without degrading local resources, and they will not be able to exist in those numbers forever. In fact, an overabundance of white-tailed deer is

Browsing white-tailed deer have created a visible browse line in this forested area after eating nearly every bit of plant material they can reach, which indicates that they have reached or exceeded their carrying capacity here.

devastating local resources. Forests in the eastern United States are heavily over-browsed, and it is difficult to find a forest that does not show a distinct browse line about six feet from the ground, which is about as high as a browsing deer can reach. The addition of new native trees, shrubs, and herbs into forest populations is near zero, because almost every tasty oak, beech, hickory, birch, viburnum, witch hazel, trillium, or Turk's cap lily that pops its head above ground is quickly eaten by deer. In most forests, the only plants surviving within reach of browsing deer are the species from Asia and Europe—the burning bush, privet, zelkova, Callery pear, barberry, garlic mustard, Japanese stiltgrass, autumn olive—as well as a few native species such as spicebush that the deer don't favor. What's more, deer now subsidize their diet with ornamental plants in suburbia (hostas and rhododendrons come to mind) that homeowners dutifully replace as soon as they are eaten. If deer continue to ravage our forests for a generation of canopy trees, the forest itself will cease to exist. The browse that deer depend on will be gone, resulting in mass starvation of any remaining animals.

But viewing carrying capacity in terms of a single species is a mistake from the perspective of conservation, because one species does not exist in isolation

Low-nesting black-throated blue warblers suffer when the plants they need for cover are overgrazed by deer.

Spring ephemerals such as squirrel corn are typical casualties of deer overabundance and overgrazing.

of other species. When one species exceeds the carrying capacity and degrades the habitat, many other species suffer as well. Economists regularly argue that the earth can support tens of billions of humans. If this were true (which it isn't), the earth could only do so at the expense of millions of other species, because water, space, food, and other resources would all be channeled toward meeting human needs. And this can occur only temporarily, because funneling a disproportionate amount of resources toward one species at the expense of others destabilizes the ecosystems that support all species. We could clarify the importance of considering carrying capacity at the community level by emphasizing that carrying capacity is the level of resource use at which diverse communities of plants and animals can be sustained without degrading the environment that supports them. We might even go so far as to say carrying capacity is the level of exploitation at which entire ecosystems can be sustained without degrading the environment.

In our white-tailed deer example, the consequences of overabundance go far beyond the impact on the deer themselves. Overbrowsing the understory of forests has changed the very structure of the forest ground layer, which, as a result, affects nearly everything that depends on a diverse community of low-growing plants. Many native communities of plants have been eliminated by overgrazing deer, including spring ephemerals such as trilliums, phlox, bloodroots, and spring beauties; understory woodies such as viburnums, dogwoods, and mountain laurels; and groundcovers like native pachysandra and mayapples. Low-nesting birds such as ovenbirds, towhees, and black-throated blue warblers that depend on these low plants for cover and food suffer as a result (Tymkiw et al. 2013; Chollet et al. 2015). Moreover, the disturbance caused by deer overbrowsing and their preference for native vegetation over plants introduced from other continents reduces the abundance of natives and encourages the rapid spread of invasive plants throughout our forests, destroying local food webs as well as water and fire regimes.

PLANTS DETERMINE CARRYING CAPACITY

The street trees of Portland remind us that our cultural perceptions can be incomplete. We tend to think of green and sustainable practices in terms of energy efficiency, recycling, and low-carbon fuels. Despite being literally green, plants are rarely included in discussions about sustainable practices, as if they didn't impact the environment. Nothing could be further from the truth, because the amount and type of plants in a landscape play a crucial role in determining the abundance and diversity of animals that can live in that landscape—in other words, the carrying capacity of that landscape. And because there are enormous differences among plants in their ability to produce the food that regulates carrying capacity, the use of the term "green industry" to label nursery owners, landscape designers, architects, and practitioners who regularly peddle low-functioning introduced species is also questionable. Through purchasing preferences, the public has insisted over the years that the green industry focus only on the decorative value of plants rather than the plants' roles as ecological backbones supporting the life around us.

It might help if we draw an analogy between carrying capacity and a personal savings account. Because the plants around you determine the carrying capacity of your landscape, think of plants as the principal in your ecological savings account—that core sum that generates the interest in your account. The larger the principal, the more interest you accumulate. If the amount of money you have as principle is large enough, you can live comfortably for years on the interest alone, as long as you don't spend down all of the interest and then dip into the principal. If you do start to reduce the principal, it will generate less interest in the future. If that happens, you will have to dip into the principal more often, and in short order, your principal will become smaller and smaller, generating less and less interest, and your savings plan will fail. Spending more money than you generate just doesn't work for very long.

This is certainly true for our ecological savings account as well. If the plants around us are the principal in our account, they are the resource base, the carrying capacity that determines how much life can be sustained. Every day they generate ecological interest in the form of food and shelter for local animals. If we increase the abundance of plants in the landscape, there will be more available food and shelter, and if we reduce the amount of plants (that is, reduce the principal in our ecological savings account), there will be less ecological interest to sustain the birds, bees, and butterflies we so enjoy. Unfortunately, if we eliminate most or all of the plants in a space, as we do every time we pave a new parking lot, build a new house, or plant a new lawn, there is no longer enough food and shelter in that space to support much of anything. So when I suggest changing our planted landscapes in ways that support other species, I am actually suggesting that we raise its carrying capacity.

GROWING NO FASTER THAN EARTH

There is one sobering caveat to our discussion of carrying capacity that I might as well put on the table. The success of Homegrown National Park, as well as any other conservation strategy, will be determined by whether or not we humans use our superior intelligence to control our numbers in socially acceptable ways before nature controls our population for us in very unpleasant ways, as she has done for all species throughout the history of life on Earth. Although we placate ourselves by frivolously labeling things as "sustainable," no approach to living on Earth can truly be sustained and no conservation strategy can succeed for long if we don't stop increasing the number of humans on the planet. Both the physical and natural resources of Earth have limits, and the human population cannot grow beyond those limits without disastrous consequences for us, other species, and the earth's ability to support life.

And yet we have already grown well beyond the earth's limits. The last time planet Earth was able to support the human enterprise sustainably—that is, provide for human needs without degrading the principle of the earth's ecological bank account—was 1975, when the earth's population was 4 billion (Juniper 2013). As I write in 2018, global population is approaching 7.5 billion, and we have seriously degraded the earth's carrying capacity—not only for humans, but for other species as well. Our goal should not be to see how many humans we can squeeze onto the planet before ecosystems collapse (our current approach), but rather to return to a population the earth can support well without stressing the natural capital on which we depend.

Controlling human numbers is not easy. To curb human numbers, we must curb human nature—an extraordinary challenge, to say the least. Without exception, we have built global economies that are based on the mythical goal of perpetual growth. This would work if the earth were growing, but it isn't and it won't, and neither will its resources. In fact, we have already degraded more than 60 percent of the earth's ability to support us (Millennium Ecosystem Assessment 2005), a statistic that should shock our personal devices right out of our hands if we would only pause for a minute to reflect on its implications for the future. And, yes, we must start thinking about the future, for it is moments away in ecological time. Quite simply, humans have exceeded and thus degraded the carrying capacity of planet Earth, and our only long-term option is to reduce our numbers to some ideal population size that the earth can sustain—not just temporarily, but, in essence, forever. We can do this by transitioning to a steady state economy that does not extract more stuff than the earth has to offer. Limiting birth rates to two children per person will temporarily cause top-heavy age classes, with more old people than young people, and this will demand the restructuring of our social safety nets (Social Security, Medicare, Medicaid, and the like) that are, in effect, Ponzi schemes that work only as long as we perpetually increase in numbers.

Although our culture denies it, perpetual growth defies the laws of physics, which, I predict, will not change to accommodate our population crisis. These challenges can be met (and are being met in Germany, Iceland, Japan, Poland, Portugal, and Italy, for example), but to do so requires that we abandon our culture of denial and start planning for futures longer than four-year political cycles. We know how to do this, and when we get serious about our future on planet Earth, we *will* do it.

CHAPTER SEVEN

Are Alien Plants Bad?

Pretty is as pretty does.

—PROVERB

THE PLANTS WE USE IN OUR LANDSCAPE determine the carrying capacity of the area. Obviously, we cannot deny the importance of soil, rainfall, altitude, latitude, sunlight, and other factors in determining the number and types of species that can thrive. Assuming the conditions required for local plant growth are met, however, the amount and type of plants in an area determine how much energy is captured from the sun and passed on to animals in higher trophic levels. An important nuance to our use of plants will determine whether we actually succeed in raising the carrying capacity of our yard. To improve a yard's ability to support life, we must use plant species that successfully create the ecological interest (in the parlance of our savings account analogy) on which other life depends. Through a twist of ecological fate, plants differ widely in their ability to do that. How, then, do we know which plants create the highest carrying capacity?

NATIVE VERSUS INTRODUCED

Nearly all of us get our plants from nurseries, but the plants in most nurseries fall into two very distinct categories: they are either native to your area—that is, they share an evolutionary history with the plant and animal communities in your ecoregion or biome—or they are plants that have developed the traits that make them unique species elsewhere. Depending on where you live, most typical garden ornamentals are native to East Asia, although nurseries in the Pacific Northwest and coastal California sell many plants from the Mediterranean region, and nurseries in the Deep South carry a plethora of tropical plants. Most landscapers don't care much about where a plant originally comes from; we choose ornamental varieties for the sole purpose of meeting specific aesthetic needs. Maybe we need a particular color or habit to complete a design, or perhaps we want a plant with a striking bloom to serve as an accent or focal point in our yard. Maybe we choose a dense evergreen to serve as a screen, or, more often than not, we choose our plants so that our yard would look just like that of our neighbors, and their neighbors, and so on. With these planting goals, the geographic origins of our plants matters not. However, if we choose plants to increase the carrying capacity of our yard, geographic origin is the first attribute we must consider. Remember that we are choosing plants to fill particular ecological roles, and plants native to the region are almost always far better at performing local ecological roles than plants introduced from somewhere else.

Novel ecosystems

For as long as *Homo sapiens* has been on the move around the globe, he has carried plants and, to a lesser extent, animals with him. Modern modes of transportation, international trade, and a keen desire to display unusual plants in our gardens have turned what was a trickle of introductions in past centuries into a torrent of new species entering North America from other lands. The influx of new species has been so great, so sudden, and so disruptive that ecologists now describe the majority of today's ecosystems as being evolutionarily novel ecosystems (Hobbs et al. 2006). They are considered novel because many of the species within them are just meeting each other for the first time in evolutionary history. This means their interactions with one another are occurring without the tempering effect of long periods of coevolution.

TERMS OF ENDEARMENT

Before we can decide on whether non-native plants are good or bad, we need to settle on some terms and specify exactly what they are good or bad at doing. Not so long ago, I was in the habit of referring to non-native plants as "alien" species. This seemed logical to me, because they were indeed foreigners imported from somewhere else. But I was taken to task for using that term, because, these days, "alien" is a pejorative term and conjures up images of helmeted creatures chasing Sigourney Weaver around a space station. Why not follow the lead of horticulturists and call non-native species "exotic" plants? That sounds nice; in fact, it sounds too nice to me and seems biased in a positive direction. Something exotic reminds me more of a beautiful belly dancer than of a plant with the potential to wreak ecological havoc in our natural areas. "Out-of-towners" has also been suggested, which is accurate enough, but at least for now, I am happy using the term "introduced" instead. After all, the ornamental plants that evolved from other locations, as well as the weeds that have plagued agriculture for centuries, were all purposefully or accidentally introduced to North America by humans.

I have also been criticized for calling particular introduced plants "bad" in some of my writings. My critics say that when we label something as bad, we are making a value judgement based more on emotion than on facts. They would have a point if I were simply making an emotional decree when labeling a plant bad. But "good" and "bad" are relative terms that can be measured; something can be considered good or bad in comparison to something else. The descriptive terms "good" and "bad" need not be based on emotion or opinion; they can be used to evaluate the relative value of two things accurately, as long as we specify what we are comparing and then measure the difference in performance between them. And that is what scientists do—they measure things!

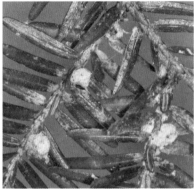

House cats kill up to four billion birds per year in the United States.

The hemlock woolly adelgid, a species native to Japan, has all but eliminated hemlock trees from our Southeastern forests.

The emerald ash borer, believed to have entered the country in wooden packing materials from China, is killing millions of ash trees.

The gypsy moth, which originated in Europe, has defoliated North American forests repeatedly for the last century.

Although some scientists are excited about the evolutionary potential of novel ecosystems (Hobbs et al. 2013), such potential will not be realized for eons, and in the meantime, novel introductions pose serious threats to the interactions that have already evolved. Many such introductions have been devastating to native populations within ecosystems, and thus to the structure and function of those ecosystems. When a novel animal predator is introduced to an area, for example, local prey often have no prior adaptations for defending themselves and quickly fall victim (Stolzenburg 2011). Consider the introduction of cats and stoats to New Zealand, which has nearly eliminated the flightless birds indigenous to that island, including the iconic kiwi. Rats that swam to islands from docked ships have destroyed thousands of sea bird breeding colonies in the ugliest of ways. And the largest extinction event in the Holocene occurred when humans colonized the Polynesian Islands some 4000 years ago (Duncan et al. 2013). Astoundingly, from 800 to 2000 species of birds were lost to human hunting and to the introduction of rats to these islands. The same type of ecological devastation has occurred time and again for both animals and plants when diseases such as chestnut blight and white pine blister rust, and non-native insects such as the hemlock woolly adelgid, emerald ash borer, and gypsy moth were introduced to North America from other lands. But the most widespread and underappreciated consequences of creating novel ecosystems are those that occur when introduced plants replace native plant communities.

Invasive species

Novel ecosystems are created by invasive species, and there are more species of invasive plants—more than 3300 in the United States alone (Qian and Ricklefs 2006)—than all of the other invasive organisms combined. Invasive plants, defined as non-native species that displace native plant communities, should not be confused with fast-growing, aggressive native plants for one simple reason: native plants, aggressive or otherwise, have been duking it out with one another, competing for space, light, water, and nutrients, for millions of years. Over the eons, native species have evolved ways to cope with one another, and the results of their interactions define the highly diverse species composition of most native plant communities all over the world. Invasive plants, in contrast, have arrived in a community within the last several hundred years, which is a blink of an evolutionary eye. They also have arrived without their suite of natural enemies—the insects, mammals, and diseases that keep them in check in their homeland. Interactions between invasive plants and our native species are anything but tried and true; invasive and native plants are just starting to negotiate what their future coexistence will look like, and it will take hundreds or thousands of generations for these negotiations to reach a compromise. Unfortunately, native plant communities are not able to negotiate

Invasive Japanese stiltgrass eliminates plant diversity on forest floors.

from a position of power, because plants that have proven to be invasive have an enormous competitive advantage over most native plant species, which enables many of them to run amuck across the landscape.

You might think images of a spotted knapweed or cheatgrass monoculture in the West, or a kudzu, privet, or phragmites (common reed) invasion in the East, would easily convince anyone that fighting such invasions is a good idea; yet questions about the wisdom of attempts to curb vegetative incursions have been raised ever since invasive ornamentals started moving from our gardens into natural areas. People concerned about the impacts of invasive plants on ecosystems have been accused of being too emotional about plant invasions, of not letting nature take its course, of ignoring the beneficial side of introduced plants, and of trying to return our ecosystems to some pristine state that has not existed for at least 14,000 years. Some criticism of efforts to control invasive plants is understandable. We spend billions each year trying to manage invaded ecosystems, and though local success is common, eradication is nigh on impossible. Even wide-scale control of invasive plants is extraordinarily difficult, particularly when these plants are for sale in nurseries. Fighting what many see as a losing battle might seem a fool's errand if there were not compelling reasons to do so. But there are good reasons to keep introduced plants

Before this wetland was invaded by phragmites, dozens of plant species thrived here; now it is a monoculture of this invasive plant

off our properties, and you can decide for yourself if they are good enough to justify the effort required to do so.

Critics of invasive control efforts concede that invaded communities are different from communities before the arrival of an invasive species, but they claim those differences are not inappropriate (Davis 2009; Davis et al. 2011). All ecosystems are in a constant state of flux, they argue. If an introduced plant enters an ecosystem and changes the diversity and abundance of native species, that change is a natural process that should not be challenged by humans—even though humans were responsible for it. They further argue that plant invasions are not bad for ecosystems because there are no records of a plant invasion causing a continent-wide extinction of a native species, and there are more plant species in North America after an invasion than before.

Are these claims true? Let's take a close look at the evidence. First, are there really more species present after an introduced plant displaces a community of native plants? The answer depends on the geographic scale being considered (Powell et al. 2013). If we have introduced 3300 new plant species to North America over the years, species diversity should now be higher. And on a continental scale, it is. But ecosystems don't function on a continental scale; they function locally, and there are oodles of studies documenting the reduction or complete elimination of one or more plant species after the arrival of an invasive plant on a local scale (such as Collier et al. 2002; Knight et al. 2007). Moreover, when counting species impacted by introduced plants, we should count not only plant species, but also the animals that eat or pollinate plants and, in turn, the predators of those animals. Every time a native plant is removed from an ecosystem, or even diminished in abundance, populations of all of the animals that depend exclusively on that plant are also removed or diminished, as are the natural enemies of those species. In sum, then, at the local scale—the scale that counts ecologically—invasive plants typically decimate local species diversity, and claims to the contrary have not been supported by rigorous field studies.

What about the assertion that invasive plants have not caused any native plant extinctions on continents? In this case, the qualifier "on continents" has to be added because the threat of extinctions caused by invasive species on islands is quite high (Downey and Richardson 2016). In contrast, on continents, native plant populations are larger and more dispersed, and there are more places for native species to escape invasive species, so even if a native population is clobbered in one place, populations may persist in other places. However, there is one biological phenomenon associated with some plant invasions that is so pernicious, even continental scales are not protecting natives from invasive species. I speak of introgression, or introgressive hybridization, where the invasive species hybridizes with a closely related native, and then through repeated backcrosses and directional gene flow, the gene pool moves closer and closer to that of the invader. This is the process by which African-ized bees have replaced the European honey bee genotypes wherever the two have come into contact in just a few generations. American bittersweet (Zaya et al. 2015) and red mulberry (Burgess and Husband 2006) are two examples

---

Oriental bittersweet, sold for its ornamental berries, has extirpated American bittersweet in many states through introgressive hybridization.

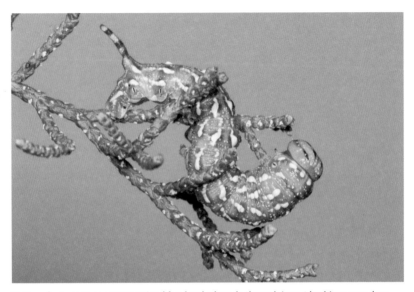

Native plants pass energy to animal food webs largely through insect herbivores such as this Doll's sphinx moth larva.

of native plants that are rapidly disappearing from their native range in a similar manner through directional introgression with oriental bittersweet and white mulberry, respectively. Both species of native plants have been replaced by their non-native congeners in all but the extreme fringes of their ranges. Unfortunately, the introgression is proceeding at such a rapid pace that the native plants' extinction seems imminent.

The claim that there are no records of extinctions caused by invasive plants may not remain true for long, though it is certainly true that the introduction of non-native plants has not created an extinction threat for most native species. That, however, does not mean ecosystem function has not been compromised at the site of each invasion. Besides, using global extinction as the only indication of harm is like saying the only symptom that warrants a visit to the doctor is death. When invasive plants such as autumn olive, buckthorn, barberry, air potato, bush honeysuckle, and phragmites invade a plant community, they replace the local native species at that site either completely, causing local extinction, or partially, causing a decline in the plants upon which that ecosystem has depended for eons. If introduced plants were the ecological equivalents of the native species they replaced, ecosystems would look different after an invasion, but they would be just as productive (though less stable). Introduced plants may, in fact, be equal to natives in their ability to produce some ecosystem services, but they pale in comparison to natives in perhaps the most critical role plants play in nature: introduced plants are poor at providing food for the animal life that stabilizes our ecosystems.

Plants, in essence, enable animals to eat sunlight. By capturing energy from the sun and storing it through photosynthesis in the carbon bonds of simple sugars and carbohydrates, plants are the basis of every terrestrial and most aquatic food webs on the planet. But animals benefit from the energy captured by photosynthesis only if they can eat the plants, or eat something that ate the plants previously. And there is the rub: insects are the animals that are best at transferring energy from plants to other animals, and, unfortunately, most insects are very fussy about which plants they eat.

The curse of specialization

Ah, how easy conservation would be if all plants delivered the same ecological benefits—that is, if all plants were the ecological equivalents of one another. We could plant eucalyptus around the world and plant-eaters everywhere would be happy as koalas. The nectar-filled butterfly bush so many people plant "to help the butterflies" would actually serve as a larval host for all butterflies (instead of only one species in southern California) and deliver up pollen and nectar to all 4000 species of native bees rather than just a few generalist bees. The resplendent quetzal would be common as starlings because it would be able to eat the fruit of all plants instead of just wild avocado. We could rename the evening primrose moth the every-plant moth, and the ornamental bamboos that are consuming yards and road shoulders would feed monarch and queen butterflies as well as milkweeds do.

Alas, specialized relationships among plants and animals are the rule rather than the exception in nature, and they are far more common than generalized relationships. This is particularly true for specialized relationships involving food webs, those interconnected relationships that transfer energy harnessed from the sun by plants to animals that eat plants, and then to other animals that eat plant-eating animals. Many people refer to this transfer of energy as a food chain, but if you were to make a diagram of a plant and include all of the species that eat that plant, as well as all of the species that eat each of those plant-eaters, the result would look far more like

Butterfly bush, an invasive shrub commonly found in many gardens, supports larval development in only one of North America's seven hundred twenty-five butterfly species.

a very complex spider web than a linear chain. Food web is a more accurate description of these interactions.

By far, the most important and abundant specialized relationships on the planet are the relationships among the insects that eat plants and the plants they eat. Most insect herbivores, some 90 percent in fact, are diet specialists—host-plant specialists that are restricted to eating one or just a few plant lineages (Bernays and Graham 1988; Forister et al. 2015). Host-plant specialization has been known to and well-documented by entomologists since the early 1960s, but scientists have never been very good at talking to one another, so the importance of host-plant specialization in gluing ecosystems together is still underappreciated by many ecologists, restoration biologists, and particularly conservation biologists. This is why proposals to reforest tropical areas of the world with eucalyptus have not been met with jaw-dropping outrage (to wit, blue gum eucalyptus from Australia is now the most abundant tree in Portugal

[McGuire 2013]), and why, more and more, shade coffee marketed as being good for the birds is being grown under the shade of non-native eucalyptus, citrus, and mango, even though these trees support little to no insect food for birds. The specialized relationships between insects and plants are so important in determining ecosystem function and local carrying capacity that it is worth spending a little time to explain why this is so, and how these relationships have come about.

Plants, of course, don't want to be eaten; they want to capture the energy from the sun and use it for their own growth and reproduction. So in an attempt to deter plant-eaters, they manufacture nasty-tasting chemicals and store them in vulnerable tissues such as leaves. These chemicals are secondary metabolic compounds that do not contribute to the primary metabolism of the plant. In other words, they are not a necessary part of the everyday jobs of living and growing. Instead, their job is to make various plant parts distasteful or downright toxic to insect herbivores. Some well-known plant defenses include the toxic compounds cyanide, nicotine, cucurbitacins, and pyrethrins; heart-stoppers such as cardiac glycosides; and digestibility inhibitors such as tannins (Tallamy 2004).

If plants are so well defended, how, then, can insects eat them without dying? This question dominated studies of plant–insect interactions for three decades, but at this point, the answer has been thoroughly delineated. Caterpillars and other immature insects are eating machines; some species increase their mass 72,000-fold by the time they reach their full size (Richards and Davies 1977). Because caterpillars necessarily ingest chemical deterrents with every bite, there is enormous selection pressure to restrict feeding to plant species they can eat without serious ill effects. Thus, a gravid (pregnant) female moth attempts to lay eggs only on plants with chemical defenses their hatchling caterpillars are able to disarm.

There are many physiological mechanisms by which caterpillars can temper plant defenses, but they all involve some combination of sequestering, excreting, and/or detoxifying defensive phytochemicals before they interfere with the caterpillar's health. Caterpillars typically come by these adaptations through thousands of generations of exposure to the plant lineage in question (Rosenthal and Berenbaum 2012). In short, by becoming host-plant specialists, insect herbivores can circumvent the defenses of a few plant species well enough to make a meal, while ignoring the rest of the plants in their ecosystem. For our

Many coffee farms grow their coffee plants under the shade of fast-growing, non-native eucalyptus under the false assumption that the trees support the insects that birds require in their diets.

purposes, however, a key point regarding host-plant specialization is that it does not happen overnight; although every once in a while an insect species coincidentally possesses enzymes that are able to disarm a plant species that it has never before encountered in its evolutionary history, it usually takes many eons for an insect to adapt to a new host plant, if it can adapt to the plant at all.

Does this mean insect specialists have won the evolutionary arms race with plants? Somewhat, but only in relation to the plant lineage on which they have specialized. When viewed across all lineages, plant defenses are very effective at deterring most insects. The monarch butterfly provides a great example. This species is a specialist on milkweeds, which use various forms of toxic cardiac glycosides to protect their tissues. Very few insects can eat plants containing cardiac glycosides, but over the ages, monarchs (in fact, the entire danaid lineage to which monarchs belong) have developed the enzymes that can make cardiac glycosides less toxic. They also have a physiological mechanism for storing these distasteful compounds in their wings and blood, rendering their own bodies unpalatable to predators. And monarchs have gone a step further: Unpleasant taste does not help a monarch if a bird has to eat it in order to discover its unpalatability. But monarchs, like many other distasteful insects, advertise their bad taste with an aposematic pattern of orange and black, which serves as a universal warning signal to would-be predators: Don't eat me. I taste bad.

Milkweeds are so named because, in addition to cardiac glycosides, they defend their tissues with a milky latex sap that jells on exposure to air. Insects that attempt to eat milkweed leaves soon find their mouthparts glued permanently shut by the sticky sap. Monarchs, however, have found a simple but amazing way to defeat this defense: they block the flow of sap to milkweed leaves (Dussourd and Eisner 1987). This is an example of a behavioral adaptation (as opposed to a physiological adaptation), and you can easily watch it in action if you grow milkweed in your yard. When a monarch caterpillar first walks onto a milkweed leaf, it usually moves to the tip of the leaf and starts to eat. If any latex sap starts to ooze from the wound, the caterpillar immediately stops eating, turns around, and crawls two-thirds of the way back up the leaf, where it chews entirely through the large midrib. That simple act severs the main latex canals that move the sap throughout the smaller leaf veins. With the canals blocked, all of the leaf tissues below (distad of) the midrib wound become latex-free, and the monarch can eat them without gumming its mouthparts. If the monarch decides to chew through most of the leaf petiole (stem) instead of

Over the eons, monarch butterflies have developed all of the adaptations required to survive and reproduce on milkweeds, but specialization has come with a cost, because milkweeds are the only plant they can eat.

the leaf midrib, latex is blocked from the entire leaf. (Incidentally, this behavior provides a convenient tool to help us find monarch caterpillars. Look for the leaf flags at the point where the monarch weakened the midrib. Any milkweed plant with a flagged leaf is or has been the home of a monarch.)

The advantage of these adaptations is obvious for the monarch, but there are also disadvantages to such specialization, especially in today's world. Unfortunately for the monarch, the ability to detoxify cardiac glycosides and block latex sap in milkweeds does not confer the ability to disarm the chemical defenses found in other plant lineages. This means that of the 2137 native plant genera in the United States, a monarch can develop (with very minor exceptions) on only one: the milkweed genus *Asclepias*. The evolutionary history of this butterfly has locked it into a dependent relationship with milkweeds, and if milkweeds disappear from the landscape, so will the monarchs. And this is exactly what has happened across the United States in recent years. Those who favor neat, lawn-lined agricultural fields, combined with an unwillingness to

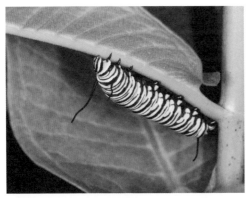

To block the flow of sticky latex sap, monarch cater-pillars chew through the midrib of milkweed leaves before they start to eat the leaf.

First it finds a suitable spot on the midrib.

It then chews through the midrib.

It chews until the midrib flags.

The caterpillar then moves to the tip of the leaf and eats with no threat from latex sap.

include milkweeds in designed landscapes, have resulted in a reduction of monarch populations by 96 percent from their numbers in the 1970s (Brower et al. 2011). Can monarchs adapt to other plant species? In theory, yes, but in reality, no. The monarch lineage has been genetically locked into a relationship with milkweeds for millions of years. Adaptation could conceivably modify this relationship very slowly over enormously long periods, but asking monarchs to suddenly (within decades) switch their dependence on milkweeds to an entirely different plant lineage—say, for example, crape myrtle—is like asking humans to develop wings. The number of genetic changes required to make such a switch reduces the probability of it happening before monarchs disappear to near zero.

Monarchs are not exceptions, either in their specialized relationship with milkweeds or in their current plight. They are typical of 90 percent of the insects that eat plants, and whose evolutionary histories have restricted their development and reproduction to the plant lineages on which they have specialized. And as we homogenize plant diversity around the world by replacing diverse native plant communities with a small palate of ornamental favorites from other lands, the insects that depend on local native species decline. We have caused these declines by the way we have designed landscapes in the past. But we can and must reverse them by the way we design landscapes in the future, for such decisions will determine how well our ecosystems function.

ECOSYSTEM FUNCTION

I talk about ecosystem function often, which usually draws blank stares. I use the term "function" because in some ways, ecosystems are like well-oiled machines, built from many interacting parts that combine to perform different functions. Creating the life support systems that keep living things, including us, fed, healthy, buffered from severe weather, and supplied with plenty of clean air and water are some of the things functioning ecosystems do every day. When we look more closely at how ecosystems function, though, my analogy breaks down. Machines have a specified number of parts, and removing some parts almost always impairs or destroys the machine's ability to function, while adding more unnecessary parts does not make the machine function better. In contrast, research over the past sixty years has shown that ecosystems are far more flexible than machines regarding the number of parts that run them. To

About 90 percent of the insects that eat plants, such as this exquisite curve-lined owlet caterpillar, can develop on only one or two plant lineages.

me, the most fascinating result of this research has been the discovery of the relationship between the number of species (parts) in an ecosystem and how well that ecosystem performs its various functions.

In 1955, at the very dawn of critical ecological thinking, Robert MacArthur published a paper in which he suggested that the stability and ability of an ecosystem to function—the ecosystem productivity—were inseparably related to the number of species residing within that ecosystem. As the number of species increased, so did both ecosystem stability and productivity. MacArthur was perhaps the most brilliant theoretical ecologist of the twentieth century, so for him this was an unusually simple hypothesis to propose. Testing it, however, would prove to be extraordinarily difficult, so he didn't. He just called this relationship the "law of nature" and left it to future generations of ecologists to test. Other hypotheses followed, including the famous rivet hypothesis proposed by Ann and Paul Ehrlich (1981) and the redundancy hypothesis by Walker (1992), but they were also largely theoretical exercises that went untested in the field. Slowly, however, as the resources required for long-term studies became available, and ecologists learned how to manipulate simplified ecosystems, direct measures of the relationship between species richness and ecosystem function began to appear in the literature (Schmitz et al. 2000; Rey Banayas et al. 2009; Reich et al. 2012; and others). Surprisingly, all of these studies were largely in agreement: MacArthur's theory had been correct after all.

Like machines, ecosystems run more smoothly, longer, and more productively when they contain all of their parts. Also like machines, some parts (species) are more vital to ecosystem function than others, and losing these parts can shut down the entire system. But unlike machines, there may be no upper limit to the number of parts that run ecosystems; every time a new species joins the ecosystem, it runs better than it did before. Unfortunately, the opposite is also true: every time we remove a species or diminish its numbers to the point where it can no longer perform its role effectively, the ecosystem becomes less productive and less stable. And that in a nutshell explains why replacing native plant communities with introduced plants compromises ecosystem function. Not only do non-native plants often reduce the number of species in the ecosystem, but they always reduce the number of interacting species. Being present but not interacting with other local species is akin to throwing a monkey wrench in a machine. The monkey wrench is a new part added to the machine, but not only is it a part that does not interact in a positive way with the other parts of the machine, it actually prevents the other parts from interacting effectively. Introduced species occupy space in an ecosystem—space that was once occupied by contributing native species—but they have not been present for the thousands of generations required to form the specialized relationships that run ecosystems.

INTERACTION DIVERSITY

Friedrich Wilhelm Heinrich Alexander von Humboldt, eighteenth-century Prussian philosopher, linguist, botanist, biogeographer, and gifted naturalist, accumulated a list of scientific contributions during his ninety years that was even longer than his name. Von Humboldt spent much of the time between 1799 and 1804 exploring the tropical regions of the Americas. Among other things, he was keenly interested in the way the natural world worked. Not only was he the first to propose that Africa and South America had once been attached to

Among his many contributions to ecology, von Humboldt first suggested that interactions among species were the driving force within ecosystems.

each other and had drifted apart, and that human activities would change the earth's climate if they continued unabated, but more to our point, he was also the first to recognize formally that it was the way species interacted with one another, and not the actual species themselves, that formed the glue holding nature together. Von Humboldt reasoned that nature was not an abstract idea, but a living, interconnected entity. The species are important, but how these species interact and the diversity of these interactions make nature a living force.

Like von Humboldt's ideas on continental drift and climate change, the importance of what we now call interaction diversity was forgotten soon after his death in 1859. What is easily apparent to most people are the species we encounter in nature, but not the myriad ways in which those species depend upon each other. A species-centric view of the world intensified as more and more species became threatened with local or global extinction in the twentieth century. And despite growing calls from ecologists to preserve entire ecosystems, what still dominates the public's perception of conservation is the plight of individual, typically charismatic, species: our inboxes are full of pleas to save the whales, elephants, rhinos, tigers, and jaguars. The importance of interacting suites of species has been too ambiguous for the public and even many scientists to appreciate.

Fortunately, this is changing, and the change is once again being led by tropical ecologists. While lamenting the local extinction of countless species throughout Central America, Dan Janzen, perhaps the most perceptive ecologist of the last fifty years, noted that the most insidious form of extinction was not the loss of individual species, but the extinction of ecological interactions (Janzen 1974). This thinking has spawned network analysis, a new way of measuring nature, and Lee Dyer, an ecologist at the University of Nevada, thinks such analyses will soon show that interaction diversity is a better predictor of ecosystem function than MacArthur's species diversity (Dyer et al. 2010). After all, Dyer argues, the interactions among species affects all aspects of ecosystems, from primary productivity, to the way populations fluctuate, to the survival and reproductive success of individual species. And in the few cases in which it has been studied, interaction diversity is devastated by the introduction of non-native plants.

I will close this discussion with some numbers, not just because they demonstrate that introduced plants reduce both species and interaction diversity, but because they hammer home how large these reductions are. A few years ago, my students Melissa Richard and Adam Mitchell set out to measure what happened to caterpillars when invasive plants created a novel ecosystem. Finding habitats that were thoroughly invaded by introduced plants such as autumn olive, multiflora rose, Callery pear, porcelain-berry, burning bush, and bush honeysuckle was easy. They typify the "natural" areas near the University of Delaware where we did our study. The trick was finding places that were still relatively free of invasive plants. Using a combination of restored sites and areas not easily accessed by deer that exacerbate the spread of invasive plants, we finally found what we were looking for: four invaded sites and four primarily native sites of similar size. We counted and weighed caterpillars at each site, once in June and

A hedgerow in Maryland invaded by autumn olive, multiflora rose, bush honeysuckle, oriental bittersweet, barberry, burning bush, and other introduced plants

again in late July. By every measure, the caterpillar community, and by extension, the community of insectivores that relied on caterpillars for food, were seriously diminished when introduced plants replaced native plants. Even with more plant biomass along the invaded transects, there were 68 percent fewer caterpillar species, 91 percent fewer caterpillars, and 96 percent less caterpillar biomass than what we recorded in native hedgerows (Richard et al. 2018). To summarize these numbers in terms of the everyday needs of the animals that eat caterpillars, we found 96 percent less food available in the invaded habitats! Interactions between caterpillars and hedgerow plants were also significantly impacted by introduced plants, with invaded hedgerows supporting 84 percent fewer interactions and 57 percent less interaction diversity than coevolved hedgerows. Had we compared the dozens of species that depend on caterpillars or their adult moths in invaded and native transects—the dipteran and hymenopteran parasitoids, as well as the assassin bugs, damsel bugs, predatory stink bugs, spiders, toads, and birds—the impact of these plant invasions on interaction richness and diversity would have been many times larger.

CONSEQUENCES

My students and I did our study in unmanaged hedgerows, which pass for natural areas where I live. Would we have seen the same impact on insect populations if we had conducted the study in a typical suburban neighborhood? The answer would depend entirely on the percentage of introduced plant life in the study landscapes. Unfortunately, in most urban, suburban, and even extra-urban landscapes, the majority of plants are not native to the area (McKinney 2002). My students and I have measured this in twenty-five-year-old suburban developments in Delaware, northeast Maryland, and southeast Pennsylvania. We didn't have to work too hard, because these landscapes contained very few plants at all—they were 92 percent lawn! But of the plants that we found, 79 percent on average were introduced from other continents. Moreover, they were largely the same species we had studied in the invaded hedgerows: Callery pear, bush honeysuckle, privet, burning bush, Oriental bittersweet, barberry, and Norway maple. Unfortunately, homeowners still landscape with invasive species throughout the country.

For years, I have speculated about the consequences of such landscaping choices for birds that many of us would like to see in our yards. I had to speculate because no one had directly measured what happens to bird populations in landscapes that favor introduced plants (although Karin Burghardt had measured this indirectly back in 2008 [Burghardt et al. 2009]). I was pretty safe in my speculations, because logic dictates that if you take away the food birds need, they won't stick around. This, as the saying goes, is not rocket science. Nevertheless, I need speculate no longer. In the first study of its kind, my student Desiree Narango measured what happens to Carolina chickadee populations and the caterpillars that support them when native plants are replaced by introduced ornamentals in suburban settings (Narango et al. 2017, 2018). For three years, Narango and a team of field assistants followed breeding chickadees in the suburbs of Washington, D.C., during the nesting season. Using video cameras at the nests, radio isotope analyses, territory mapping, vegetation analyses, foraging observations, and citizen scientists, she was able to quantify all of the variables required to model population growth of chickadees as a function of the percentage of introduced plants within the chickadee's breeding territories.

Desiree found far more than I have space to describe, but here are some of the highlights of her research. Throughout her study, parent birds foraged for food on native plants 86 percent of the time. Compared to primarily native landscapes in her suburban study sites, yards dominated by introduced plants produced 75 percent less caterpillar biomass and were 60 percent less likely to have breeding chickadees at all. Apparently, chickadees were able to assess the quality of the landscape before they decided whether or not to set up house in one of Desiree's chickadee boxes. If a chickadee did build a nest in a yard with many introduced plants, it contained 1.5 fewer eggs than nests in yards dominated by natives, and those nests were 29 percent less likely to survive. Chickadees that decided to nest where there were not enough caterpillars fed their chicks spiders and small homopterans such as aphids, but these food items did not compensate nutritionally for the lack of caterpillars in their chicks' diets. Nests in yards dominated by introduced plants produced 1.2 fewer chicks and slowed chick maturation by 1.5 days compared to nests located in yards with lots of native plants.

Some of these differences may not sound very substantial, but cumulatively they are making a huge and negative impact on suburban chickadee populations. The consequence of these differences was that chickadee populations achieved replacement rate—that is, produced enough chicks each year to replace the adults lost to old age and predation—only in yards with less than 30 percent introduced plants. Unfortunately for the chickadees in Washington suburbs, Narango found that, on average, 56 percent of the plants are introduced.

Although these results are not surprising, they remove the guesswork from understanding how much our plant choices impact the life around us. Here is solid evidence that, at least for Carolina chickadees, introduced plants are not the ecological equivalents of the native plants they replace, and it is hard to imagine why other insectivorous birds would not be similarly affected by introduced plants. Desiree's research helps us understand that the plants we have in our yards make or break bird reproduction, not the seeds and suet we so dutifully buy for our feathered friends, although those supplements certainly help birds

Most suburban yards are devoid of native plants, which means they provide little or nothing in the way of food and shelter for birds and other wild creatures.

Although it depends on seeds much of the year, a Carolina chickadee must find thousands of caterpillars to rear one clutch of young.

after they have successfully reproduced (Marzluff 2014). Her results also give us insight into what is happening beyond our yards in the once natural areas that have been invaded by ornamental plants from Asia and Europe. We can now better understand one of the factors that have caused 432 species of birds to decline at a perilous rate in North America (North American Bird Conservation Initiative 2016). Our studies are not the only ones that show how seriously insect populations are affected by introduced plants. Dozens of other studies have found the same effect over and over again. This study and others to come should supply all the motivation we need to help create Homegrown National Park.

DO BIRDS CARE IF A BERRY IS NATIVE OR NOT?

Some studies have focused on whether berries produced by introduced plants made up for the loss of insects in sustaining birds. Mark Davis, a botanist at Macalester College in St. Paul, Minnesota, asked this interesting question in 2011, when the controversy over fighting invasive plants was at its peak. He concluded, quite logically I might add, that if birds readily eat berries from introduced plants, they must not care about the evolutionary origin of those berries. As anyone who has witnessed a flock of cedar waxwings strip a European crabapple of its fruit in minutes can attest, many birds do eat berries produced by introduced plants. In fact, the birds are why many introduced plants are so highly invasive: they eat seed-laden fruit, fly some distance away, and poop out the seeds that will start another generation of that species. So if introduced plants are providing lots of good food for birds, maybe they aren't so bad after all?

The heart of this question focuses on the nutritional value of berries produced by different species of shrubs. We can assume that all berries are equal in what they deliver to birds, but assuming things is not science. Science is hypothesis testing, and that is the focus of Susan Smith and her collaborators at the Rochester Institute of Technology. They've been testing the hypothesis that berries are nutritionally equivalent whether they are produced by native shrubs or introduced shrubs (Smith et al. 2013). The study focuses on berries produced in the fall for two reasons: nearly all of our invasive shrubs produce their berries in the fall (I can't think of any that don't), and both migrating and overwintering birds depend on fall berries for the fats they need either to fuel their migration or to build fat reserves for the long winter months if they don't

The red-eyed vireo, along with many other birds, are the primary dispersers of seeds within berries.

Berries differ a great deal in their nutritional value: berries produced in the summer are high in sugar, while berries such as spicebush fruit that are produced in the fall are high in fat.

migrate. Smith's work has revealed a surprising and distressing pattern: berries from introduced Eurasian plants such as autumn olive, glossy buckthorn, bush and Japanese honeysuckle, and multiflora rose contain very little fat, typically less than 1 percent, while berries from natives such as Virginia creeper, wax myrtle, arrowwood viburnum, spicebush, poison ivy, and gray dogwood are loaded with valuable fat, often nearly 50 percent by weight. There is variation among species, but the pattern is clear: berries in the introduced plants studied so far are high in sugar at the time of year when our birds need to consume berries high in fats.

Why would a bird eat a berry that does not meet its nutritional needs? I can think of two reasons. First, when an invasive shrub moves into a habitat, it typically eliminates the native shrubs that used to grow at that site. One consequence of such thorough invasions is that the berries produced by the invasive shrub are now the only ones available, and with no other options, hungry birds eat them. I can suggest a second reason birds might eat nutritionally bereft berries by asking this question: why do you eat a sugar-coated donut when you ought to be eating a garden salad? Despite having the most developed reasoning capacities in the animal kingdom, we humans are unable to resist eating what is not good for us—so why do we expect birds to be able to bypass a sugary treat? Despite this logic, new research suggests that birds do, in fact, care whether a berry is native or not, and they discriminate against introduced berries whenever they have the option. Studies by both Smith (2015) and Yushi Ogushi (2017) have found that when fall migrants stop to rest and eat in a habitat loaded with invasive shrubs, they do not stay long. Instead, they linger in habitats with plenty of the spicebush and arrowwood viburnum berries they need to fuel their migration.

Why would shrubs from Asia make high-sugar berries in the fall when it seems that selection should have favored high-fat berries in the fall? Every time I talk about Smith's findings, someone asks me that question. Unfortunately, we don't know nearly enough to answer it with any confidence. Are the Asian invasives displaying the phenology of berry production they evolved in Asia, or did the move to North America change their phenology? What are the migration patterns of birds in Asia, and have they exerted selection on Asian shrubs the same way North American birds have selected for fall berries high in fat? These are interesting questions that I hope someone will pursue soon.

Like many birds, downy woodpeckers need to consume fall berries high in fat, such as poison ivy berries, to help them make it through the winter or to give them the energy required for migration.

COSTS VERSUS BENEFITS

Every now and then a paper is published that points out the ecological benefits delivered by introduced plants. The conclusion drawn by the authors is always the same: if introduced plants are doing good things in local ecosystems, perhaps we should tolerate, or maybe even encourage, their presence. This logic has been used to justify planting more eucalyptus in California and to discourage municipalities from spending money to fight invasive species almost everywhere. I would agree with this line of thinking on one condition: the net effect of the plant in question must be positive. Benefits cannot be viewed in ecological isolation; they must be compared to the costs or disadvantages associated with a particular plant as well. Demonstrating the benefits that a plant delivers is meaningless unless we also measure the ecological costs that plant brings to the system. Only in this way can we estimate whether the benefits outweigh the

Silver-spotted skipper larvae have the physiological adaptations that enable them to use kudzu as a host plant, but that does not mean that kudzu is good for the environment.

costs, or vice versa. If the net effect of an introduced plant improves ecosystem function, then yes indeed, let's rethink our bias against it.

Here is a typical scenario. Kudzu, the Asian plant that now exclusively occupies more than seven million acres in the Southeast (Forseth and Innis 2004), serves as a host plant to our native silver-spotted skipper butterfly. This is not just a rumor; I have taken images of silver-spotted skipper caterpillars hiding within and eating the curled leaves of kudzu plants in Mississippi. The skipper, a legume specialist, has found the defensive chemicals of kudzu to be within the range of its detoxification abilities, and so it can reach maturity on the otherwise nutritious leaves of this introduced species. We can put the new host association between the sliver-spotted skipper and kudzu in the benefits column; kudzu is creating a food web option that did not exist before its introduction. Sticking to the food web theme, though, we now have to consider whether kudzu is having any negative impacts on local food webs. The answer, of course, is yes. When kudzu smothers young oak trees in Camden County, Georgia, for example, a host option for 454 species of caterpillars disappears under mounds of kudzu leaves. Similarly, if black cherry is eliminated from this kudzu patch, 324 caterpillar species are lost. If willows, hickories, and maples are covered by kudzu,

247, 229, and 223 species of caterpillars are lost, respectively (National Wildlife Foundation 2015–2018). Such losses will hold for all of the woody plant genera lost to kudzu at this single site in Georgia. What about herbaceous plants? Well, if kudzu smothers goldenrod, as it surely would, 94 species of caterpillars are lost. If it covers native asters, another 80 species are lost. Sunflowers lost to kudzu host 67 species of caterpillars, horsenettle 67 more species, and so on. By allowing kudzu to invade an area in Camden County, Georgia, we have gained a host plant for silver-spotted skipper, but we've lost host opportunities for the caterpillars of more than 1000 other moths and butterflies. And, as with all food webs, removing food from the base of the web reverberates throughout the entire web, impacting all of the species that eat the caterpillars lost to kudzu. No doubt, the net effect of this kudzu invasion is not just slightly negative; it is hugely negative in terms of supporting local biodiversity.

WHY NOT LET NATURE TAKE ITS COURSE?

Some might actually use our kudzu example as evidence that nature is, in fact, taking its course to bring introduced plants into functional relationships with their new ecosystems. After all, the silver-spotted skipper has already adapted to kudzu (or, more likely, already possessed the necessary physiological adaptations required to eat kudzu when kudzu first arrived in Georgia). Isn't this evidence that nature is repairing itself without our help? In the long view, I suppose it is; unfortunately, however, the rate at which nature is repairing the damage we have inflicted is so incrementally slow compared to the rate at which we keep inflicting damage that true repair will not occur fast enough to prevent the loss of what we now know as nature.

Author and food activist Michael Pollan once asked whether there should be a statute of limitations on being alien (1994). In other words, if a plant or animal has been in North America long enough, shouldn't we consider it ecologically equal to the organisms that evolved here? It's a good question, but it reflects how difficult it is for most people to comprehend the immense periods of time required to build evolutionary relationships. In my view, "native" is not a label a species earns after a given period of time. It is a term that describes function. For example, a plant should be considered a native when it acts like a native—that is, when it has achieved the same ecological productivity that it

In some parts of White Clay Creek State Park in Newark, Delaware, introduced plant invaders make up more than a third of the vegetation.

After nearly 500 years in residence in North America, *Phragmites australis*, the common reed, supports only 3 percent of the insects it supports in its European homeland.

had in its evolutionary homeland, when it has accumulated the same number of specialized relationships that had been nurtured by the native plants it displaced, and when it has accumulated the same number of diseases, predators, and parasites that species that evolved in North America must endure. Time in residence is not the variable to be measured here; it is the rate at which local organisms adapt to the plant's presence.

The common reed, *Phragmites australis*, provides a great example. The European genotype that has displaced wetland vegetation from the Atlantic coast to the shores of Lake Michigan has inhabited North America for hundreds of years. There is good evidence that it was used as packing material in the holds of the earliest sailing ships some 500 years ago. In Europe, phragmites supports 170 species of insects. After hundreds of years of residence in North America, only 5 insect species are using phragmites as a nutritional resource (Tewksbury et al. 2002). Adaptation is happening, but at a glacial rate typical of evolutionary change. And phragmites is not an exception. Very slow rates of adaptation have been recorded for a number of introduced plants. The paperbark tree, *Melaleuca quinquenervia*, for example, has been growing in the Florida Everglades for more than 130 years. In Australia where it evolved, it supports 409 species of insects; in Florida, it supports only 8 North American insects (Costello et al. 1995). And in California, only 1 species has adapted to eucalyptus after 110 years (Strong et al. 1984), and no species have adapted to the prickly pear cactus, *Opuntia ficus-indica*, after 260 years (Annecke and Moran 1978). In short, it takes enormous periods of time before introduced plants act like the natives they replace.

There are other reasons why it's not a good idea to "let nature take its course" after introducing hundreds of non-native plants. First, these types of introductions are anything but natural. We have perpetrated a biological exchange of species across the globe so rapidly—instantaneously on an evolutionary time scale—that it is a phenomenon ecosystems have never before encountered in the history of life on earth. There is no natural response to counter our meddling. Moreover, by moving introduced plants beyond the suite of natural enemies—the insects, mammals, and diseases—that keep them in check in their homelands, we have stacked the competitive deck against native plants that do have to contend with hundreds of herbivores and diseases. Expecting native species to duke it out successfully with introduced plants is ecologically unrealistic. And that is why natural succession from one type of plant community to another is essentially dead when invasive plants enter the scene. Disturbances these days more often than not do not progress from grassland to meadow to scrub to forest; instead, the land becomes frozen in a perpetual tangle of invasive vines and shrubs. At least, that has been the case during the last thirty years. Will our native species prevail in the end? Some think they will, and I hope those optimists are right. But I wonder.

To those who claim it's a fool's errand to try to restore nature to some mythical pristine state, I say, "That's not the goal!" At least that's not my goal. The natural world that existed before humans entered North America at least 14,000 years ago is gone, along with the Pleistocene megafauna that shaped it. The modified natural world that Europeans encountered in the fifteenth century is gone, along with the Native Americans who shaped it. Since then, we have dramatically altered our watersheds, soils, forests, wetlands, and grasslands to meet our agricultural needs and then our industrial needs. And we continue to change the land today. But none of these changes means we have to destroy or can afford to destroy ecosystem function. Wherever and whenever we can, we must reassemble the coevolved relationships between plants and animals and among animals themselves that enable ecosystems to produce the life support systems we all need. Introduced plants only hinder our efforts to so.

—

This has been a heady and heavy discussion, but it needed to be to cover this complex situation. At this point, enough studies have been completed and enough information gathered that we can now definitively answer the question, Are alien plants bad? The preponderance of evidence says yes. Compared to native plant communities, introduced plants are bad at supporting insects and are thus bad at supporting insectivores. They are bad at supporting specialist pollinators, complex food webs, stable food webs, local biodiversity, interaction diversity, and, most important of all, they are bad at supporting ecosystem function. I recognize there are social constraints on what we can do with this information, but we should no longer accept the notion that introduced plants are the ecological equivalents of the native plants they replace.

Our environmental boat has sprung a leak. Many of us are trying to repair the leak; others are bailing to keep us afloat until the leak is plugged. What is baffling, though, is that far too many of us are dumping new buckets of water into our boat, as if sinking it will not be a problem for them. At this point, each of us must decide what role we will play in the future: Will you be a bailer or a dumper? Your choice of plants in your yard will determine what role you have chosen.

CHAPTER EIGHT

Restoring Insects, the Little Things That Run the World

If all mankind were to disappear, the world would regenerate back to the rich state of equilibrium that existed ten thousand years ago. If insects were to vanish, the environment would collapse into chaos.

—E. O. WILSON

IMAGINE A WORLD where a good fairy has the power to bestow life in exchange for the ability to live peacefully among its inhabitants. This is more than a fair exchange, for in Fairyland, everyone depends on the fairy. There are no political debates about whether people can afford to protect the fairy's well-being; indeed, no one could exist without her. She does not demand it, yet many worship the fairy and pray she will never withhold her gift of life. Sermons are written about her benevolence, cities are built in her name, global holidays are held in her honor, and those who threaten her are punished convincingly. The fairy must be sustained at all costs, for without her, everyone is doomed.

Now let's turn Fairyland into reality. Planet Earth is identical to Fairyland, with one difference: life as we know it is not sustained by a miraculous fairy, but by insects. E. O. Wilson called insects "the little things that run the world" because of the many essential ecological roles they play every day (1987). Insects pollinate 87.5 percent of all plants, and 90 percent of all flowering plants (Ollerton et al. 2011), and plants turn energy from the sun into the food that we and an unimaginable diversity of birds, mammals, reptiles, amphibians, and freshwater fishes need to exist. Insects are also the primary means by which the food created by plants is delivered to animals. Most vertebrates do not eat plants directly; far more often, they eat insects that have converted plant sugars and carbohydrates into the vital proteins and fats that fuel complex food webs.

Insects, then, sustain the earth's ecosystems by sustaining the plants and animals that run those ecosystems. And the more plants and animals, the better. As you've learned, ecosystems with many interacting species are more stable, more productive, and better able to support huge human populations than depauperate ecosystems with few species. Insects also provide much of the planet's pest control in the form of millions of species of predators and parasitoids that keep food webs in balance. Insects rapidly decompose dead plants, releasing the nutrients they contain for use by new plant life. And by keeping the planet well-vegetated, insects maintain the watersheds in which we all live, keeping our water clean and minimizing the frequency and severity of floods. As if all of that were not enough, the plants that insects pollinate sequester enormous amounts of carbon within their bodies and within the soil around their roots, carbon that would otherwise be in the atmosphere, wreaking havoc on the earth's climate.

Just as no one could exist without the fairy in Fairyland, humans would last only a few months if insects were to disappear from Earth (Wilson 1987). It is remarkable, then, that our cultural relationship with insects is not one of awe and appreciation, but one of disgust and animosity. We have created a culture in which insects and their arthropod relatives are maligned. In the name of protecting crops and fighting a few disease vectors, we have declared war on all insects and we kill them whenever we can. We have sponsored National Insect Killing Week (Weis 2009) and taught our children to fear every insect they see, rather than respect those few that might sting to defend their nests.

We are winning our undeclared war against insects at our own peril. Precipitous declines in populations of the European honey bee, the 4000 species of bees native to North America, and beautiful butterflies such as the monarch and Karner blue have gotten our attention, but many other insects are disappearing utterly without notice. We have already driven three North American species of bumblebees to extinction (Pearson 2015), and in Europe, about 30 percent of the grasshoppers, crickets, and katydids are facing extinction (Burton 2017). Flying insects in Germany have declined in abundance and diversity 79 percent since 1989, and 46 species of butterflies and moths have disappeared from German soil altogether (Hallman et al. 2017). Similar statistics are coming to light in England and other parts of Europe. By killing insects, we are biting the hands that feed us, and that has led to the most alarming statistic of all: invertebrate abundance (the number of insects) has been reduced 45 percent globally since 1974 (Dirzo et al. 2014).

As insane as our war against insects may seem, and as effective as it has been, I am nevertheless optimistic that we can form a new relationship with insects and treat them like the good fairies they are. Why am I willing to put a smiley face on this? Two reasons: First, our response to insects is, in part, an oversimplified, innate reaction to things that hurt or annoy us now and in our distant past. If bees sting and mosquitoes bite, it is easy to group all small flying things into one category we label enemy. But we now have knowledge on our side, and with that, we have the capacity to be a bit more discriminating. We can easily learn to distinguish the good from the bad, the helpful from the irritating. By numbers, nearly all insects are harmless and beneficial, and we can learn to appreciate rather than kill the insects that we rely on just as easily as we can learn that dogs, though genetically identical to wolves, can be our loving pets rather than our predators.

I am also confident that we can form a new relationship with insects, because we have done it many times before with other creatures. Whales still swim in the sea because we have learned to value the earth's largest species as majestic living beings instead of as mere providers of lamp oil. Wolves hunt in Yellowstone again because we now know they are essential to the long-term persistence of the greater Yellowstone ecosystem. Puffins, auklets, and murres have returned to many Aleutian Islands because we have decided not to let the rats we introduced to those islands eat their brains as they sleep. Fish breed again in the ocean sea kelp beds off California because we now prefer that our sea otters keep sea urchins in check rather than become our fur coats. Egrets fly once more over the Everglades because we value their showy feathers more in flight than on ladies' hats. And we are well on our way to forming a new relationship with bees, even though they sting, because we have finally realized we cannot exist without their pollination services. These and many more examples of how we have come to appreciate rather than destroy nature have convinced me that we can also learn to share the earth with the most essential of all creatures, the little insect fairies at our feet.

PRIORITY ONE: RESTORE THE PLANTS

Thank goodness the things that run the world are little. If we needed to share our neighborhoods with big things like tigers, elephants, bison, and giraffes, we would be challenged, indeed. Not only is it easy to create a world in which insects can coexist with humans, it is easy to create landscapes in which they actually flourish. All we need to do is include more of the right kinds of plants in our landscape designs.

Plants, of course, are the foundation of all terrestrial food webs. Their presence in our landscapes is not optional, but I fear even this most basic ecological fact is unknown to the general public. I may be overthinking this, or maybe my monothematic mind has finally gotten the better of me, but while watching one of the *Star Wars* movies with two of my grandkids, it struck me that George Lucas didn't seem to recognize that you could not have thriving populations of humans—or wookies, or vulptexes, or fathiers, or porgs, or any other living thing—without a source of energy. Most of the planets featured in his movies are deserts devoid of plants, and one even boasted of being entirely covered by a continuous city. Plants that convert our sun's (or any sun's) energy into food are not even an afterthought in that galaxy far away. Now before you *Star Wars* fans unleash The Force against me, I do get it that this is fantasy. Nevertheless, these movies cannot help but convey the subliminal message that we do not need the support of a plant-based ecosystem to exist. Maybe not in fantasy galaxies, but on planet Earth, plants are essential, and the sooner our city planners and developers fully embrace this reality, the better off we all will be.

Now that we agree that to sustain insects and other animals, we need the plants that support them, we can decide which plants we need most. But to do that, we should choose which insects we want to keep. There are a lot of insect species in the world—three to four million by most estimates, more than 164,000 of which can be found in the United States alone. As you might expect, they do lots of things and live in almost every niche in every environment, save

Although they look, act, and function much like true caterpillars, sawfly larvae are actually the immature stages of a group of wasps. **BELOW** Animal life simply would not exist without plants.

Antarctica. What they all have in common is that they are directly or indirectly tied to plants, either by eating some part of a living plant, by existing solely on dead plant tissues such as fallen leaves or rotting logs, by developing on the muck dead plants create when they fall into water, by eating another insect that has developed directly on plants, by eating an insect that has eaten an insect that has developed on plants, by being a parasite on a mammal that has eaten plants, and so on. You get the picture: insects need plants. But the class Insecta is so large and so broad that trying to create habitat for all of its members in a given space is not only impractical, but impossible. I suggest we narrow our focus a bit. So let's focus on enhancing populations of the two groups of insects that arguably have the greatest impact on terrestrial ecosystems: those that contribute the most energy to local food webs—that is, the insects that are larger, more numerous, more edible, and more nutritious than most other insects—and those responsible for most of the pollination required by plants. I speak of caterpillars, the larval stages of moths, butterflies, and the group of Hymenopterans we call sawflies, plus the 4000 species of bees native to North America.

SPOTLIGHT ON CATERPILLARS

The early bird catches the worm. True enough, if you consider caterpillars to be worms (which they're not). Caterpillars are the mainstay of most bird diets in North America, particularly when birds are rearing their young. Very few bird species, including American robins, regularly feed their babies earthworms. If we are going to landscape in a way that builds the populations of a particular group of insects, what makes caterpillars a good target? For one thing, there are many types to work with: estimates of the number of species of Lepidoptera in North America top 14,000. This is fewer than the number of beetle species in North America (25,000), but unlike most caterpillars, beetles are difficult prey items to find and eat and therefore do not contribute as much to local food webs. Beetles spend most of their lives hidden underground, within seed pods, or tunneled deep in wood. They also sport much thicker exoskeletons, particularly as adults, and often have spiny, stiff legs, which make them difficult for creatures like birds to eat and digest. Caterpillars, in contrast, are typically exposed on vegetation, and their exoskeleton is thin and flexible, making most of a caterpillar digestible food instead of indigestible chitin. Their bodies are

Parent birds use their beaks to stuff insects down their offspring's throat like a plunger.

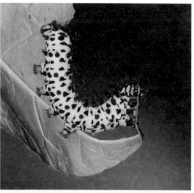

Caterpillars, such as this calleta silkmoth larva, are soft bags of easily digested nutrients for birds.

More like a little tank than an accessible food item, a scarab beetle has a thick, indigestible exoskeleton.

Caterpillars like this Glover's silkmoth are unusually nutritious sources of food, particularly for rapidly growing nestlings.

like soft bags filled with food, and this may be one of the caterpillar's most important attributes. If you have ever watched a bird feed its nestling, you know it is not always a gentle process; many birds forcibly stuff the food item down the nestling's throat using their beak as a plunger. Insects with sharp edges can injure delicate little chicks during such feeding bouts.

Most caterpillars are relatively large compared to many other kinds of numerous insects. It takes 200 aphids, for example, to equal the weight of

Most terrestrial birds in North America feed their nestlings caterpillars far more often than any other type of insect.

one medium-sized caterpillar. If you are a bird looking for insects, would you choose to hunt and handle 200 aphids, or would you seek out a single caterpillar to feed your babies? Finally, caterpillars are more nutritious than most other insects. They are high in protein and fats, and they are the best source of healthy carotenoids for birds (Eeva et al. 2010), particularly during the breeding season when few high-carotenoid berries are available.

What do carotenoids have to do with good nutrition? Plenty. In fact, carotenoids are essential components of a healthy diet. They stimulate immune systems, improve color vision and sperm vitality, and serve as antioxidants that protect proteins and DNA from oxidative damage. In birds, carotenoids provide yet another benefit: they are a major component of colorful feather pigments. Brightly colored feathers are a signal of good health, and in many species, females identify strong mates based on the color vibrancy of male feathers. Some carotenoids, such as beta carotene from carrots, lycopene from tomatoes, and lutein from kale, are familiar to us; others, such as alpha carotene, crocetin, and zeaxanthin, are less known. Familiar or not, we all need carotenoids, yet none of us can make our own. We and all other vertebrates (those creatures like humans and birds that have a backbone) must get our carotenoids directly or indirectly from plants. If you are a bird that does not eat plant tissues, you have to get your carotenoids by eating something that did eat plants.

Here's the key: for reasons we don't yet understand, caterpillars contain more than twice as many carotenoids as other insects and more than three times as many as spiders (Eeva et al. 2010). It follows, then, that if birds need carotenoids to raise healthy young (they do), and if caterpillars provide the best and most easily obtained source of carotenoids for birds during the breeding season, then caterpillars may not be optional components of breeding bird diets; it is instead likely they are essential to successful reproduction. As with all parts of nature, there are exceptions to this generality. A few bird lineages such as finches, doves, and crossbills can rear young on a milky substance they make from seeds, and raptors for the most part feed their young mammals, fish, or other birds. But most of North America's terrestrial bird species, some 96 percent in fact, rear their young on insects rather than seeds and berries (Peterson 1980), and we are learning that in most of those species, the majority of those insects are caterpillars or adult moths. Caterpillars are so important to breeding birds that many species may not be able to breed at all in habitats that do not contain enough caterpillars (Narango et al. 2018).

How many caterpillars are enough? That of course depends on which bird species we are talking about. There are enough caterpillars in a habitat when parent birds can find caterpillars fast enough to enable three to six nestlings to grow from eggs to slightly larger than adults in under two weeks for most cup nesters, and a little longer for cavity nesters. Birds have an astonishingly fast growth rate. What's the rush? Of all the perilous times in a bird's short life, the nesting period is the most dangerous period of all. Baby birds are sitting ducks, as it were, vulnerable to dozens of types of predators. Their best defense is to minimize their exposure to such danger by minimizing their time in the

Even though it is a member of the thrush family and is related to robins, the eastern bluebird relies more on caterpillars than any other food source while feeding its young.

Because cavity nesters like this tufted titmouse enjoy more protection from predators than open cup nesters, they can rear their young at a more leisurely pace.

nest. Cavity nesters enjoy better protection from predators, so they can afford to grow a bit more slowly, but most species still fledge after only sixteen days in the nest.

To achieve such rates of growth, nestlings must eat often. You and I eat three or four times a day (maybe five if we include snacks). A typical nestling, in comparison, eats a full meal thirty to forty times a day! That means a parent (or couple) raising five chicks must bring food to the nest about 150 times a day. They are busy indeed! Most birds forage primarily within a well-defined territory surrounding the nest. For Carolina chickadees, this is about 160 feet in all directions from the nest, or an area approximately two acres in size. Birds are constrained to foraging so close to the nest because it is imperative that they conserve time and energy. If a bird had to fly two miles each time it searched for food for its chicks, it would not be able to make the necessary 150 trips each day without wearing itself to a nub. The point here is that nesting territories must contain lots of food concentrated in a relatively small area or the nests will fail.

So how many caterpillars is enough? Few people have actually sat down and counted all of the prey items birds bring to their nests, although today's video technology is making this easier and easier (until you have to go through each frame of the video and record the data). But those who have had the patience to do this have recorded astounding figures. Robert Stewart, for example, made detailed records of a Wilson's warbler pair while they were feeding their young in 1973. He found substantial differences in how hard the male and female parents worked at this endeavor. The male was no slouch, carrying food to the nest 241 times in a single day, but the female put him to shame, because on that same day she fed the nestlings 571 times. This rate was maintained over the five days he watched the nest. Stewart did not count the actual number of caterpillars the pair brought to the nest; feeding was rapid and often a parent carried more than one caterpillar in its beak at a time. Yet even if only one caterpillar was brought to the nest each trip, the pair would have brought in 812 caterpillars per day, or 4060 caterpillars in the five days Stewart watched the nest. The chicks he observed stayed in the nest only eight days before they fledged.

These observations are not exceptional. Field researchers have watched bobolinks bring food to their nests 840 times a day for ten days in a row (Martin 1971). Sapsuckers feed their nestlings 4260 times, downy woodpeckers 4095 times, and

We know that birds must bring thousands of caterpillars to their nest while rearing young because of research on birds such as the Wilson's warbler, bobolink, downy woodpecker, and hairy woodpecker.

Even after chicks leave the nest, parent chickadees continue to feed them for another twenty-one days.

hairy woodpeckers 2325 times (De Kiriline Lawrence 1967). All of these species regularly bring in multiple prey items per trip. Feeding rates among European birds are also high. Single-day feeding rates for ten different species of passerines averaged 259 trips to the nest (Bussman 1933). Perhaps the most complete records of feeding rates were made by Richard Brewer, who in 1961 counted the caterpillars that Carolina chickadees brought to their nests throughout the nesting period. Brewer found feeding rates of 350 to 570 caterpillars per day, depending on the number of chicks in the nest. Over the course of a typical nesting period (sixteen days on average), chickadee parents delivered 6000 to 9000 caterpillars to bring one nest of tiny (three ounce) birds to fledging. But a parent's job is not over when the chicks leave the nest. Carolina chickadee parents, for example, continue to feed their young for up to twenty-one days after fledging. No one knows how many additional caterpillars are required before young chickadees no longer depend on their parents for food.

Now let's think about an ideal neighborhood that contains not just one pair of one bird species, but thriving populations of many species. I, for one, am greedy when it comes to enjoying nature; I want Carolina chickadees in my yard, but I also want cardinals, titmice, blue jays, and Carolina wrens. I want red-bellied and downy woodpeckers, white-breasted nuthatches, yellow warblers, Kentucky warblers, robins, wood thrushes, ovenbirds, and indigo buntings. And I am so greedy that I also want bluebirds, catbirds, common yellowthroats, great crested flycatchers, yellow-billed cuckoos, mockingbirds, eastern kingbirds, field sparrows, chipping sparrows, and grasshopper sparrows. If each pair of all of these species requires thousands of caterpillars to breed successfully, imagine the number of caterpillars Cindy and I would need our yard to produce to provide food for all of these birds and their young! I cannot imagine it, and I think about this all the time!

Which plants are best for caterpillars?

It is clear that if we want to landscape a yard to accommodate as many caterpillars as possible, we need to use plants that serve as hosts for the most caterpillar species. But which plants are those? Assembling this information is not a trivial task. There are some 2137 native plant genera in the lower forty-eight states, and most of them contain species that serve as host plants for one or more species of caterpillars. Records of these host associations have been made over the past century by naturalists, ecologists, and particularly by Lepidoptera taxonomists, and these are scattered in their writings throughout thousands of papers and books. Needless to say, finding and categorizing this information requires a combination of old-fashioned library work plus the handy search tools of the digital age.

But that's just the beginning. All of the information in these records has to be regionalized. A caterpillar species that eats persimmons in Union County, New Jersey, for example, may not occur in St. Louis County, Missouri, even though persimmons grow in that Midwestern county. We need to create a match between the plants and caterpillars that occur in each county of each state, which is another monumental task of data manipulation, with 3007 counties in the conterminous United States. Fortunately, my steadfast research assistant, Kimberley Shropshire, who has helped me with all aspects of my research since 1992, was up to this task. With financial support from the U.S. Forest Service, she created a mammoth database in a little more than a year, which has become the basis of a search tool developed by the National Wildlife Federation, the Native Plant Finder (look for it at http://www.nwf.org/NativePlantFinder). She has ranked plant genera that occur in every county of the United States in terms of their ability to host caterpillars. Now, simply by entering your postal code, you can find out which woody and herbaceous plant genera native to your area are best at serving as host plants for caterpillars. Audubon has created a similar website, Plants for Birds (at https://www.audubon.org/plantsforbirds), based on her work. These two sites have removed one of the biggest obstacles to homeowner restorations: we no longer have to wonder what plants we should add to our landscapes.

Keystone plants

In addition to providing a valuable resource for people nationwide who are interested in raising the carrying capacity of their property, Shropshire's work revealed a striking pattern: wherever one looks—be it in north, south, east, or west, or the plains, deserts, forests, or mountains—just a few plant genera are providing sustenance for most of the Lepidoptera so important to our food webs. We knew from our previous work in the Mid-Atlantic states that not only were native plants far superior to introduced species in their ability to sustain caterpillars, but native plants themselves varied by orders of magnitude in their ability to host caterpillars (Tallamy and Shropshire 2009). We know that some genera, such as *Quercus* (oak), *Prunus* (cherry), and *Salix* (willow), host hundreds of caterpillar species, while for others, such as *Cladrastis* (yellowwood) and *Empetrum* (crowberry), there are no records at all of caterpillars using them. This is interesting in itself, but when Shropshire assembled data for each county, we saw that this pattern held everywhere and we could quantify it: wherever we looked, about 5 percent of the local plant genera hosted 70 to 75 percent of the local Lepidoptera species!

I refer to these hyperproductive plants as keystone plants, because they so closely fit the meaning of Robert Paine's classical terminology (1969). While studying predator–prey interactions in West Coast tidal pools, Paine found that keystone species had a disproportionately large effect on the abundance and diversity of other species in an ecosystem. He likened such species to keystones, because, like the center stone in an ancient Roman arch supports the other stones that make up the arch, keystone species support other species in their ecosystem and help them coexist. Remove the keystone and the arch, or ecosystem, falls down. Keystone plants are unique components of local food webs that are essential to the participation of most other taxa in those food webs. Without keystone plants, the food web all but falls apart. And without some minimal number of keystone genera in a landscape, the diversity and abundance of the many insectivores—the birds and bats, for example, that depend on caterpillars and moths for food—are predicted to suffer.

Cardinal

Titmouse

Nuthatch

Catbird

Common yellowthroat warbler

Indigo bunting

Chipping sparrow

White eyed vireo

Bluebird

It is hard to imagine how many insects are required to support thriving populations of the cardinals, titmice, nuthatches, indigo buntings, catbirds, common yellowthroat warblers, chipping sparrows, bluebirds, and white eyed vireos that breed in our yard every year.

The implications of this phenomenon for homeowners, land managers, restoration ecologists, and conservation biologists are enormous: to create the most productive landscapes possible—that is, landscapes in which the plant matter provides for the largest number of edible insects—we have to include species that belong to keystone plants. This is a nuanced but incredibly important extension of our knowledge about how native plants contribute to ecosystem function. Before discovering the existence of keystone plants, we overestimated the degree to which most native plants contribute to food webs and assumed that if a plant was native, it contributed a lot. We now know that a few native genera contribute much more than most others, and we cannot ignore them if we are to produce complex, stable food webs. A landscape without keystone genera will support 70 to 75 percent fewer caterpillar species than a landscape with keystone genera, even though the keystone-less landscape may contain 95 percent of the native plant genera in the area (Narango et al. 2018). This runs contrary to the age-old maxim that the more diverse a planting is, the more productive it will be. On one level, this is certainly true, because a diverse plant community will support more caterpillar species than a monoculture. But now we know that to be richly productive, plant communities must contain at least some keystone plants.

Let it be an oak

I don't remember the day I decided that *Quercus alba*, the white oak, was my favorite tree, nor do I remember why I thought so. I was an impressionable preteen very much into tree-climbing, so maybe I was simply excited by the majestic spread of a mature field-grown white oak with branches low enough to tackle safely. Or it could have been the immense size or great age often attained by these trees—superlatives impressed me then, as they do now. Serendipitous or not, I had no idea when I boldly declared the oak to be the king of deciduous trees that fifty-five years later, my research would show that, in many respects, I was right: oaks are ranked number one among temperate zone species in several measures of performance.

Derived from the Celtic *quer*, meaning fine, and *cuez*, meaning tree, *Quercus* species are fine trees indeed, with hundreds of species globally (taxonomists argue about the exact number, with estimates ranging from 400 to 600 species). More than 90 species of oaks occur in the United States and often dominate all

White oaks and their relatives are the very best trees you can plant in your yard for wildlife in 84 percent of U.S. counties.

forest ecosystems in North America except the great coniferous forests of the North and the driest deserts of the Southwest. Ecologically, oaks are superior plants, and it would be easy to make a convincing case that they deliver more ecosystem services than any other tree genus. Many species are massive and sequester tons of carbon in their wood and roots, and they pump tons more into the soil. They are long-lived as well, with some species achieving 900 years, including periods of growth, stasis, and decline. Thus, the carbon they pull from the atmosphere is locked within their tissues for nearly 1000 years.

In many ecosystems, oaks are also superior at stalling rainfall's rush to the sea. Their huge canopies break the force of pounding rain before it can compact soil, and their massive root systems, some extending more than three times the width of the canopy from the main trunk, prevent soil erosion and create underground channels that encourage rainwater infiltration instead of runoff. Lignin-rich oak leaves are slow to break down once they fall from the tree, and they create excellent resilient leaf litter habitat for hundreds of species of soil arthropods, nematodes, and other invertebrates. For me, though, all of these contributions to ecosystem function pale before the contribution oaks make to food webs.

Our early work showed that oaks in the Mid-Atlantic region supported hundreds of caterpillar species—557 to be exact (Tallamy and Shropshire 2009)—and at least 934 species nationwide, making oaks by far the best plants to include at home if you want to support food webs. If you think of a plant as a bird feeder, which is exactly what it is, then in most regions, the oak makes the most food. To put this level of productivity in perspective, most other common trees in the Eastern deciduous forest are slackers in comparison. Tulip poplar (*Liriodendron tulipifera*), for example, supports only 21 caterpillar species, black gum (*Nyssa sylvatica*) supports 26, sycamores (*Planatus occidentalis*) support 45, persimmon (*Diospyros virginiana*) supports 46, hemlock (*Tsuga canadensis*) supports 92, and sweetgum (*Liquidambar styraciflua*) supports 35. Like oaks, native willows and cherries are also highly productive, but they surpass oaks only in a few counties. In fact, oaks are ranked either number one or two in their support of the food web in 84 percent of all U.S. counties in which they occur (Narango, Tallamy, and Shropshire 2018).

For those who study plant–insect interactions, determining why the genus *Quercus* supports so many more species than other plant genera is a worthy

The spun glass slug caterpillar is one of the 557 species of caterpillars in the Mid-Atlantic states that develops successfully on oaks.

undertaking, and a number of hypotheses have been advanced (Southwood 1972; Lewinsohn et al. 2005; Agrawal et al. 2006; Becerra 2007, 2015; Condon et al. 2008; Grandez-Rios et al. 2015). Perhaps the large size or geographic range of the genus contributes to the number of caterpillar species supported. Or maybe the degree to which oaks are related to other North American plant genera—that is, their phylogenetic isolation or lack thereof—could help determine their productivity. Or perhaps their ecological apparency—their abundance, great physical size, and extended lifespan—in the landscape is a factor. Equally suspect is the type of chemical defense employed by oaks, which protects their leaves primarily with tannins that may be easier for insect herbivores to circumvent. The most likely explanation of all is that each of these factors contributes in some way to oak caterpillar productivity. But we don't need to understand precisely why oaks help food webs better than other plants; we just need to know that they do and that we should use them accordingly in our landscapes and restorations.

Most moths drop to the ground when it's time to pupate and either tunnel into the soil like the sphinx moth revealed in the photo on top, or spin a cocoon in the leaf litter like the luna moth above.

Some moths complete their lifecycles on their host tree, such as this polyphemus moth that built a cocoon, but this is unusual.

Completing the lifecycle

Regardless of which trees, shrubs, or perennials we employ to increase the abundance and diversity of caterpillars on our properties, we have to use these plants in our landscapes in ways that enable the caterpillars they support to complete their lifecycles. Caterpillars undergo complete metamorphosis, a type of development that comprises four distinct stages: egg, larva, pupa, and adult. What's relevant here is that for most caterpillar species, only two of these life stages, the egg and larval stages, are completed on the host plant. Most

CLOCKWISE Native pachysandra, woodland phlox, foamflower, and wild ginger

You can build a layered landscape that's perfect for pupating caterpillars using shrubs such as the piedmont azalea (*Rhododendron canescens*).

caterpillars crawl off their host plant before molting to their pupal stage. Oaks in Chester County, Pennsylvania, for example, serve as hosts for 511 species of caterpillars. A few of these, such as the polyphemus moth, spin their cocoon on the host tree itself after they have eaten their fill of oak leaves. But 480 species, some 94 percent, fall to the ground when the caterpillar is fully grown, where they either burrow into the soil to pupate underground or spin a cocoon in the leaf litter under the tree. Groundcovers planted under trees, such as native pachysandra, woodland phlox, foamflower, ginger, or native shrubs, make perfect sites for moths to complete their development safely.

The exodus a caterpillar makes from its host plant before it pupates is undertaken by more than just oak specialists. Monarch caterpillars almost never form their chrysalises on milkweeds; they crawl off to other structures, often yards away from the milkweed plants on which they developed, and many monarch watchers wonder if the caterpillars have been eaten by birds. The pipevine

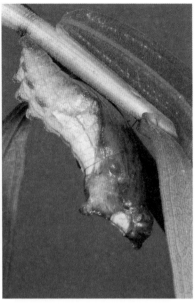

Although a monarch caterpillar hatches, feeds, and grows on milkweed, it often travels yards away from its plant host to form its chrysalis on another plant.

This pipevine swallowtail forms its chrysalis on goldenrod.

swallowtails in our yard really take a hike when they're fully grown. I have found their chrysalises halfway up our oak trees and attached to the side of our house, and one was even hanging from one of our picture frames in our living room! Some of these caterpillars have crawled more than twenty-five yards from their pipevine hosts. Experts think the evolutionary motivation for such trekking is for caterpillars to put distance between themselves and their host plant before they enter the defenseless pupal stage. Lepidopteran pupae are prey for not only hungry birds; they are also targets for numerous species of other predators and parasitoids, many of which go to a host plant to search for food. The longer a caterpillar stays on its host, the greater the chances it will be discovered and attacked by one of its enemies. By crawling some distance from its host plant before it pupates, the caterpillar has decentralized the effective search zone for the predator. Instead of caterpillar predators being able to search just a few square feet for their prey, they must search thousands of square feet, a low return task that is usually not worth the time or energy involved.

This massive suburban oak produces thousands of caterpillars each year, but the hard-packed lawn beneath it does not provide adequate pupation sites for them.

This survival mechanism is very effective at reducing mortality in the pupal stage, but it forces us to think beyond the needs of caterpillars when we landscape. Not only do we have to provide food for developing caterpillars, we also must provide the microhabitats their pupae require to survive. Your yard may include an oak tree that can feed hundreds of caterpillars, but more often than not, that oak will be surrounded by mowed lawn growing in compacted soil. When your caterpillars drop from the tree, they will find no leaf litter in which to spin their cocoons, because each year you neaten up after the leaves fall. If a caterpillar species tunnels into the soil, it will have to search far and wide to find soil loose enough to permit burrowing, and the longer it has to search, the greater the odds of it being pulverized by your lawn service or squashed on your driveway or street. These challenges are even greater in urban environments, where trees are often surrounded by cement.

Fortunately, providing safe pupation sites in our landscapes is not an insurmountable problem. In fact, it can be a new and satisfying gardening goal. We can meet the needs of our caterpillars by replacing some of our lawn with three-dimensional plantings. Annual or perennial beds, spring ephemeral showcases, groundcovers, or shrub plantings of regionally appropriate species—any portion of your landscape that is not regularly trammeled by feet

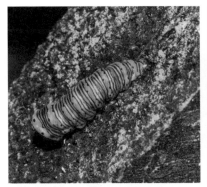

The beautiful wood nymph, here halfway through chewing its way into a moss-covered log, is among several caterpillar species that pupate within rotting wood.

Pupation sites such as logs or rocks can be artfully added to your garden.

and lawnmower wheels will quickly develop a thick O horizon, a soil layer with a high percentage of loose organic matter, that is perfect for pupating Lepidoptera, as long as you don't rake away that black gold each year during your spring and fall cleanups. In 2014, Rick Darke and I authored *The Living Landscape: Designing for Beauty and Biodiversity in the Home Garden* (Timber Press). In that book, we devoted substantial space to regional charts that provide plant options for different landscape uses. If you are looking for ideas about how to add plants to your ground layers, this is a great place to start. Pocket prairies in the Midwest, pollinator gardens in the East, and the rocks and root masses of plants such as silver sage, bush poppy, creosote bush, and rabbitbrush in Southwest xeriscapes all provide excellent places for caterpillars to pupate out of harm's way. To the surprise of many, caterpillar species such as the beautiful wood nymph, the greater oak dagger moth, and Harris's three spot tunnel into soft wood to pupate; adding an artfully placed decaying log to your garden enables such species to complete their development.

We need to take one final safety precaution to help our caterpillars complete their lifecycles. Once they emerge as adult moths from their pupae, they have to survive long enough to find a mate and then long enough for the females to locate their host plant and lay eggs. A few species, such as the white-marked tussock moth, accomplish this in just a few hours, but most require several days to a few weeks. During this time, species with mouthparts need to eat to maintain their energy, and most of the time they are eating nectar from

A WEEK WITH WAGS

It bugs me when people call insects bugs, not just because the term is taxonomically incorrect (only suborder Heteroptera insects are true bugs), and not because it diminishes the diversity and importance of insects. It bugs me because those who derogatorily lump all creatures with six legs into one group do so because they don't know one insect from another. And if they don't know one insect from another, they won't care about any of them. We humans distinguish objects and beings from one another by naming them. Learning names is usually step one toward

David L. Wagner, caterpillar guru extraordinaire

becoming more familiar with something or someone, and familiarity is necessary before we can develop empathy—before we can care whether it lives or dies, flourishes or goes extinct. When we see a dog or cat, horse or possum, we don't say, "oh, a mammal." We recognize their differences because we long ago learned to distinguish them by name. None of us has close friends whose names we don't know. Yet, here I am, asking you to develop a caring, nurturing, protective relationship with insects before you can even name them. Fortunately, the tools needed to do just that are becoming more and more available.

I recently had the chance to spend a week with David "Wags" Wagner collecting caterpillars in southern Arizona. Wags is a longstanding faculty member at the University of Connecticut, and he is the foremost national authority on the larvae of macro-Lepidoptera (we'll just call them caterpillars). Although he is best known among his peers for his scientific publications, Wags also has a growing reputation among laymen because of his two books, *The Caterpillars of Eastern North America* (2005) and *Owlet Caterpillars of Eastern North America* (2012). Currently he is tackling the mammoth job of assembling material for *The Caterpillars of Western North America*. These books are elaborately illustrated with images of caterpillars that you and I can, and do, encounter in our yards.

Without attempting to, Wags has become one of the unsung heroes of biodiversity conservation. When I told him this, he looked puzzled. But here's my reasoning: Before Wags wrote books to help us identify these insects, when we found a caterpillar, we had little hope of learning what it was. If we didn't know its name, what it looked like as an adult moth or butterfly, what its preferred host plants were, or some other piece of natural history to grab our attention and make this creature memorable, there was little chance we could make an emotional connection with it. And if

The white-blotched heterocampa is one of the fascinating caterpillars we can look for in our yard if we have an oak tree.

we didn't care about the caterpillar, there was no hope of mounting a conservation effort on the caterpillar's behalf. We would not learn how vital the caterpillar was to local food webs; we would not come to appreciate the value of including the caterpillar's host plants in our landscapes; and we certainly couldn't appreciate that the caterpillars in our yards were the most important and accessible currency by which the complexity and stability of our local food webs could be measured.

Whether Wags agrees with me or not, I believe this is his gift to conservation. His work has rescued us from an endless cycle of ignorance by making caterpillars accessible. Now, any one of us can flip through his books and match what we have found with a species. In minutes, we can learn that our caterpillar is the gray-edged bomolocha and that it is a specialist on black walnut; or it is the turbulent phosphila, a specialist on greenbrier. We can marvel at the strange and beautiful form of the curve-lined owlet pictured in his book and make it a personal goal to find this elegant caterpillar ourselves someday. We can learn that the hickory horned devil we just found is declining rapidly as a species, yet it is reproducing right in our yard. And we can learn that the white-blotched heterocampa developing on our white oak requires oak leaves, and if we cut that tree down we will lose that beautiful caterpillar.

With Wags's books in hand, we can form a lasting relationship with a previously unknown world of biological diversity. Just as Robert Tory Peterson introduced us to the world of birds, Wags has given us the tools with which to connect to, sustain, and maybe even love caterpillars, one of the most important forms of organic diversity on Earth. To watch him labor daily from 8 a.m. until 2 a.m. to bring us the caterpillars of western North America was truly inspirational.

Home security lights are deadly to nocturnal insects.

nocturnal flowers. A 2017 study by Eva Knop at the University of Bern suggests that our propensity to light up the night sky is not helping moths in this regard. Using night-vision goggles, Knop counted insects that visit flowers in areas with no artificial light. When she added lights to those same areas, she found that when the lights were on, insect visits (think moths, as they are the primary flower visitors at night) declined 62 percent. Either the moths simply avoided spaces that were well-lit, or they were fatally attracted to the lights as if the lights were Sirens.

Don't ask why insects are so drawn to lights, because two centuries of research have not produced a satisfying answer to this puzzling question. The point is that insects do fly into light sources by the millions, and each night, we light up nearly the entire world to their detriment. Lights reduce insect populations in several ways. A light can kill an insect directly after it repeatedly collides with the bulb. Or the frenetic flight about the bulb can fatally exhaust an insect by burning up its energy reserves. Insects that don't beat themselves to death or die of exhaustion are nevertheless waylaid from their normal activities of seeking host plants or mates. Night lights waste precious time in the short adult lives of insects. Finally, an insect drawn to a light becomes an easy target for hunting bats, or, if an insect is sitting near a light, it's game for arthropod predators such as daddy long-legs, carabid beetles, hanging scorpionflies, damsel bugs, ants, assassin bugs, mantids, and spiders. Those insects that are still alive at dawn are often picked off by birds that quickly learn that lights are an easy place to find breakfast. Whether you use lights at night so you can see your way around your property or to discourage intruders, consider putting motion sensors on each one. That way, instead of illuminating your property all night every night, the lights turn on only when you, or an intruder, are out and about in your yard. That simple act will make an enormous difference for insect populations, particularly for night-flying moths and, thus, for caterpillars.

RESTORING NATIVE BEES

The second group of insects we should be sure to subsidize in our landscapes is native bees, not because bees are important components of food webs (they are not; other than a number of specialized parasitoids and a few specialized birds, very few animals eat bees), but rather because, as pollinators, they maintain a diverse base for terrestrial food webs, our plants. Though many types of insects are credited with important roles in pollination, including moths, butterflies, beetles, wasps, flies, and ants, the members of the Apoidea, bees, perform the lion's share of pollination duties.

When you think of bees, an image of the domesticated European honey bee may come to mind. Honey bees were brought to North America with the earliest colonists because they could be managed and deployed where they were most needed, and because they were generalist pollinators that were so good at pollinating the many Old World crop plants the colonists also brought with them. Before the colonists imported honey bees, however, all of the animal pollination in North America (13 percent of our plants are wind-pollinated) was accomplished by native pollinators, primarily by nearly 4000 species of native bees.

Like so many of our insects, honey bees are in trouble. In the past, beekeepers expected to lose 10 to 15 percent of their colonies each winter, but dramatic, and in some cases sudden, declines in honey bee populations became apparent across the country and, indeed, globally between the years 2003 and 2007. A suite of ills—from mites, to viruses and bacteria, to abusive pollinating demands—has been blamed for these declines, now referred to as colony collapse disorder. The threat to agriculture from the loss of honey bees is so obvious that even our politicians have noticed, and today, saving pollinators has become a politically correct mantra. The press is full of articles telling us

Some think that any insect that visits flowers is a pollinator, but in reality, butterflies like this Canadian tiger swallowtail do very little pollination compared to bees.

Sweat bees are important North American native pollinators.

Bumblebees are another important group of native pollinators.

Honey bees have suffered colony collapse disorder from a series of human-induced insults.

that we have to save bees because they pollinate a third of our crops. This is true enough, but as a sole motivation for saving bees, this reasoning is short sighted in the extreme. There is an even more compelling reason, beyond maintaining a diversity of fruits and vegetables in our supermarkets, to save pollinators from local or global extinction. Pollinators are essential to life as we know it on planet Earth. In addition to pollinating a third of our crops, animals (bees, bats, hummingbirds, and others, but mostly bees) are responsible for pollinating 87 percent of all plants and 90 percent of all angiosperms (flowering plants) (Ollerton et al. 2011). So if pollinators were to disappear, 87 to 90 percent of the plants on planet Earth would also disappear. Not only would such a loss be a fatal blow to humans, it would take most other multicellular species with it as well.

One positive result of colony collapse disorder in European honey bees is that it has drawn long-overdue attention to our North American bee species. Many of our nation's top entomologists, including Sydney Cameron (University of Illinois), Sam Droege (Patuxent Wildlife Research Center), Claire Kremen (UC Berkeley), Rachael Winfree (Rutgers), Marla Spivak (University of Minnesota),

Jarrod Fowler (Xerces Society), and Doug Landis (Michigan State University), are scrambling to understand the current plight and conservation of native bees in human-dominated landscapes. Although they have studied only a handful of our native bee species, they have found most to be in steep decline. Half of the Midwest's native bee species have disappeared from their historic ranges in the last century (Burkle et al. 2013). Four species of bumblebees have declined

WHAT IS A POLLINATOR?

It is logical to assume that all animals that go to flowers for pollen and/or nectar actually pollinate those flowers. But the opposite is true; most animals that go to flowers do not end up pollinating the flowers, even if they successfully remove pollen and nectar from those flowers. It is more accurate to call these animals flower visitors and reserve the term "pollinator" for animals that successfully transfer pollen from the male stamens to the female pistils of flowers. Butterflies, for example, get lots of credit for being great pollinators because they spend so much time nectaring at flowers. But this credit is not deserved; most butterflies take from flowers without giving back much in return. Butterflies do not have a body shape conducive to transferring pollen for most flowers. Even bees that have specialized adaptations for pollinating a particular flower genus may visit other flower genera without transferring any pollen. Because pollen and nectar are costly for flowers to produce, many flower genera have developed elaborate shapes, such as extremely long corolla tubes, very narrow corollas, or closed petals, that make access to their nectar difficult. The evolutionary idea in these cases is preventing generalist pollinators from taking the pollen and allowing only specialist pollinators access to the pollen because they are more likely to deliver it to another flower of the same species. Specialized interactions between flowers and their pollinators are largely responsible for the myriad sizes and shapes of flowers and bees in nature.

96 percent just in the last twenty years (Cameron et al. 2011), and 25 percent of our bumblebee species are at risk of extinction (Williams and Jepsen 2014). It is not a stretch to assume that the thousands of species not yet examined are similarly challenged in today's world of glyphosate-ready corn and soybeans, lawns, and deadly roads. It is clear that we must all act quickly to save our pollinators, but to do this effectively, we have to understand who our pollinators are and what they need to thrive in our yards.

To make supporting pollinators at home part of mainstream culture, we must leap over an educational hurdle. Unfortunately, the thought of thriving bee populations in our yards is too often a non-starter. After all, where there are bees, there are bee stings—or so many people think. I have been asked to recommend native plants for school yards, but invariably they request plants that would not attract bees—ergo, plants that do not have flowers. Ignorance reigns supreme when it comes to bees, so let's set the record straight.

Bees do sting, or at least the females do, but only in self-defense or in defense of their hive. Males have no stingers, which are actually modified egg-laying devices. This is important point number one. Out of all 4000 species of native bees, only bumblebees, a mere 46 species, have what we might call hives, though they are tiny compared to honey bee hives. The rest of the species are solitary and never aggressively defend a home space. When a person is stung by a bee, the perpetrator is nearly always a honey bee that has been stepped on with bare feet (that was my first sting as a toddler) or that is defending its hive. Important point number two is that, while foraging at flowers, bees are not aggressive at all. They are focused solely on gathering as much pollen and nectar as possible. You can prove this to yourself by petting the next bee you see at a flower. The bee might fly off, but it won't sting you. The passive nature of foraging bees means that we can walk among flowers crawling with bees with no fear of being stung.

Before you send me an angry e-mail listing all the times you have been stung at home, read on. The most common misconception about the source of painful stings stems from mistaken taxonomy. People frequently mistake yellowjackets for bees. Yellowjackets are predatory wasps. They are not bees and they are not pollinators. But like honey bees, they are social species that construct large hives in the ground or in trees, which they defend as aggressively as they can. I include bald-faced hornets in this group; even though they are black and white instead of yellow and black, bald-faced hornets are close relatives

Bumblebees are one of the few groups of native bees that are social and create small nests of daughters.

Many people mistake aggressive and stinging yellowjackets for bees.

of yellowjackets and behave just like them. Unlike honey bees, each individual yellowjacket or hornet can sting repeatedly, making any close encounter an unpleasant one. The good news is that there is no national movement to encourage people to harbor yellowjacket nests in their yards. The solitary bees with which we need to share our yards and parks are harmless.

Making bees feel at home

Like most creatures, native bees need the basics to exist: shelter, food, and water. Let's deal with living quarters first. About 70 percent of our native bee species nest in the ground, while others nest within wood or pithy plant stems or in any nook or cranny of the appropriate size. Bumblebees, our only native bees that are always social, favor shallow holes in the ground that are protected from rain. Abandoned mouse nests within a rock wall are ideal real estate for bumblebees, because they are protected from rain and also from digging predators such as possums, raccoons, and foxes that love to eat bumblebee larvae. If you don't happen to have an abandoned mouse nest or a rock wall, you can simulate a nest by burying a full roll of toilet paper about three-quarters of its length in the ground (center hole facing up) in a site totally protected from rain or runoff. To protect the roll from rain, you can build a small, three-sided wooden house with an inch-wide hole drilled in one side. Bumblebee queens

You can invite bumblebees to build nests in your yard by burying a toilet paper roll; a queen will chew a hole into the roll and start her colony in early spring.

A bee in the genus *Colletes* guards the entrance to her nest.

fly around each spring evaluating every hole they find as a potential nest site. Chances are good that a queen will enter your makeshift nest box and chew her way into its center to set up a cozy nest.

Ground-nesting bees are easy to accommodate, as long as the soil is loose enough for bees to excavate—that includes almost all uncompacted soil types, but not hard-packed clay. Ground nesters prefer bare patches of dry soil with a slight southern slope, so if your yard has such spaces, avoid walking on them because that compacts the soil beyond use for the bees. We are not talking about huge areas—two square feet of bare soil can provide housing for a number of reproducing female bees. And do not use lawn fertilizer near nests, because lawn products usually contain pesticides that do not make life easier for our bees. Species of native bees, particularly in the families Colletidae, Halictidae, Andrenidae, and some Apidae such as long-horned bees and squash bees, will construct nests in the ground all season long, but their nests are most evident in early spring before they are obscured by vegetation. Look for holes in the ground surrounded by small mounds of excavated soil. In an active nest, you'll spot a bee leaving or entering the hole every few minutes. The holes lead to tunnels several inches deep with side shafts here and there, each containing a ball of pollen and a developing bee larva. On any site on your property that is ideal for ground-nesting bees, many individuals may nest within that small

Leafcutter bees like this one can nest in any hole of the appropriate diameter.

space. This can result in lots of bees frenetically coming and going, a spectacle that may be a bit scary at first, but these bees are busy rearing their young and will do their very best to ignore you. You can sit and watch them for hours with no ill effects!

Pithy stem nesters are a bit more particular in where they nest. Stems of many herbaceous plants such as goldenrods, blackberries, giant ragweeds, and native hydrangeas are essentially hollow except for a loose fibrous material that bees can easily remove. These cavities make perfect nesting sites. Mason bees, small carpenter bees, and small resin bees will tunnel into the stem, remove the pith from a section several inches long, and then construct a sequence of cells starting at the end of the cavity farthest from the entrance hole. Each cell is packed with pollen, upon which the bee lays a single egg. She then seals off that particular cell and starts to provision the neighboring cell. In this way, the stem contains several developing larvae at once, each spread a day or two apart from neighboring larvae. When development is complete, each larva pupates within the cell. If it is early summer, the young adult will emerge from its cell by chewing a hole directly to the outside, where it begins its search for a mate to start its own family. If a larva matures at the end of the season, the resulting prepupa will stay within its cell all winter, complete its development early in the spring, and emerge as an adult as soon as flowering plants are in bloom.

Woody stem nesters such as carpenter bees and some species of mason bees behave almost identically to pithy stem nesters, except they build their nests in soft wood rather than soft stems. Soft wood could be found in a downed log or branch, or just as often in a dead branch still attached to a tree. Dead elderberry branches are good examples of suitable nest sites for these bees because they are so easily excavated. Finally, species such as mason bees, yellow-faced bees, and leafcutter bees often choose existing cavities for their nests. I cannot tell you how many times I have found the snout of my watering can or even my outdoor water spigot plugged with leafcutter bee nests.

In terms of bee conservation, there is a common theme here. Bees cannot nest or overwinter in our yards unless we provide what they need to do so. Most people do have open patches of ground, plants with pithy stems, easily excavated wood, and nooks and crannies on their properties, but many of us work hard to eliminate these valuable resources. Our fall cleanup is particularly hard on bee populations; the senescing stems of black-eyed Susans, penstemons, sunflowers, and all of the other perennials we are so anxious to cut back after they have bloomed are where pithy stem nesters are hoping to spend the winter. Similarly, that dead elderberry branch we feel compelled to prune off and the large elm branch that fell during a summer thunderstorm are now homes for bees that favor soft wood. The social edict to neaten up is often in direct conflict with the needs of our native bees. There are opportunities for compromise, however, if we think about it for a few minutes. Maybe we can gently cut off

Not only do elderberry shrubs provide great forage for pollinators when in flower, but dead elderberry stems are ideal nesting sites for native bees because their wood is soft and easily excavated.

goldenrod stems near the ground, but rather than mulching them, we can tie them together like a decorative bundle of corn stalks and stand them up for the winter somewhere out of public view. The bees and katydid eggs within should be able to make it through the winter just fine.

Meeting the needs of our specialists

After we have provided housing for our bees, we need to feed them—that is, we need to meet the nutritional needs of both adult bees and their developing larvae. Adult bees eat pollen and nectar, while larval bees develop exclusively on pollen. That seems simple enough, but we need to pay attention to some essential particulars here. First, we need to think about timing. Bees, like most other multicellular organisms, must eat every day. Because the pollen and nectar they need comes from flowers, we need to ensure that plants are blooming in our landscape throughout the season. Bee communities are active most of the year in most parts of the country. Even in colder regions such as New England, native bee species are on the wing from March through October (Fowler 2016). The need for a continuous sequence of flowering plants in our landscapes is not a trivial challenge. It requires plant choices that are choreographed to flower one after another throughout those months. Although having more than one species of flowering plant blooming at once is desirable and gives bees nutritional options, a landscape that goes through a two- or three-week period with no available blooms is deadly to bees.

Blooming phenology is only one thing to consider when we're landscaping for bees. We also have to recognize that many bee species require pollen from particular plant genera in order to reproduce. Like caterpillars, many bee species have become host-plant specialists over evolutionary time. In fact, in the Mid-Atlantic region, nearly 30 percent of the native bees are host-plant specialists (Fowler 2016). It's no wonder that natural selection has favored bee specialization; plants differ from one another in so many ways, some bees can meet their nutritional needs more easily if they develop the specialized adaptations that enable them to find, gather, transport, and digest the pollen of particular plants in an efficient way. Think about it: plants differ in when they flower, how long they flower, and the sizes, shapes, and colors of their flowers. They also differ in their pollen morphology and the types of amino acids, lipids, proteins, starches, sterols, and chemical defenses their pollen contains. Some

BEE HOTEL

In recent years, we have discovered how easy it is to attract many species of stem- and wood-nesting bees, as well as species that nest in nooks and crannies, using commercial or homemade bee hotels. There is scarcely a garden-related conference or trade show these days that does not offer a variety of bee hotels for sale. Hang one in a dry space in your yard at the end of winter and in short order the bees will be busy rearing their young inside. The number of bees you attract is often a simple function of the size of the bee hotel.

Large bee hotels may attract lots of bees, but they also invite bee predators.

Bee hotels are the perfect solution to increasing the nesting capacity in our yards—or so we all thought. What should have been obvious from the start is that by concentrating nesting opportunities in one place (in one large bee hotel), we have made it very convenient for bee predators, parasitoids, and diseases to wipe out our bees (MacIvor and Packer 2015). Find one bee and you have found them all! This doesn't mean bee hotels are useless, but it does mean we need to make them much smaller, with just a few cells each, and scatter many throughout our yards. In this way, we are not putting all of our bees in one basket. If a bee enemy finds a hotel, it will not have easy access to all the bees on your property.

Create and scatter several small hotels like this one about your property to make it more difficult for bee enemies to find their prey.

Various species of goldenrod meet the needs of many specialist bees as well as local generalists.

generalist bees are good at dealing with all of this variation and can do well on a variety of pollen species. But others become finely tuned to the characteristics of particular plant genera, so if we don't include those genera in our plantings, those specialist bees cannot survive in our yards.

How do we support both our specialists and our generalists? Sam Droege of the Patuxent Wildlife Research Center in Maryland says that we should meet the needs of our specialists, and the generalists will follow. For example, in the Mid-Atlantic region, fifteen species of bees can rear their larvae only on the pollen of goldenrod. And thirteen more species require pollen from asters, eleven need evening primrose (*Oenothera* spp.), eleven more cannot reproduce without willow pollen, and nine depend on blueberry pollen (Fowler 2016). If you are trying to help native bees but you plant butterfly bush, xenias, and impatiens, you will see generalist bees and you may bask in a false sense of accomplishment, but without goldenrods, asters, evening primrose, blueberries, and native willows, sixty-nine species of specialist bees that would have been able to use your yard as a refuge will be absent. Clearly, saving our bee specialists by planting what they have specialized on is the key to saving diverse bee communities around the country.

CHAPTER NINE

What Have Weeds Done for Us Lately?

The greatest scientific discovery was the discovery of ignorance.

—YUVAL NOAH HARARI

MY WIFE IS THE WEEDER IN TALLAMYLAND. Like so many of us, she enjoys orderly landscapes and prefers that our plants grow only in their designated spaces. Yet she also understands that the animals in our yard depend on our native plant communities, and she worries about their fate when she weeds particular plants out. Her sensitivity has grown over the years as we have learned more and more about how our various plants interact with other creatures. For example, the other day she came in from the yard and said, "I know goldenrod supports a lot of species, but what specifically would suffer if I weeded some out?"

"Well, let's see" I said. "Goldenrod leaves support 110 species of caterpillars in Southeast Pennsylvania and many species of leaf beetles and June beetles. In our area, its flowers provide pollen and nectar for 35 bee species, 15 of which use only goldenrod pollen; myriad wasps; as well as long-horned, scarab, blister, phalacrid, ripiphorid, and ladybird beetles. And don't forget that goldenrod nectar is an important source of energy for migrating monarchs. Goldenrod flowers are favorite hunting sites for crab spiders and praying mantids, and its seeds feed a number of wintering sparrows, juncos, and finches—and birds use them to line their nests in the spring. Its stems provide housing for native bees during both summer and winter and support four species of gallers [insects that create galls] as well as several stem-boring caterpillars, and they are feeding sites for many plant and leafhoppers. Who knows how many species goldenrod roots support."

"Hmm," she said, "I will leave as much as I can." Before returning to the garden, she looked back at me and said, "I wish more people knew this. You have to tell them."

Good idea!

MARKETING ISSUES

A weed is defined as a plant out of place. The problem with that definition, though, is that it is entirely subjective. To an ecologist, any plant that has no evolutionary history in a given space is a plant out of place, and that includes the beautiful ornamentals we have introduced from elsewhere. But to most gardeners, any plant that tries to grow within a design that did not specify it is a plant out of place, and that includes all of the native plants that had grown on that site for thousands of years. Our subjective perception of where plants belong is how so many of our native plants came to be called weeds. When Europeans first arrived on the shores of what is now Virginia, they imposed European farming techniques on the New World. This was not the way Native Americans grew crops, but Europeans planted monocultures and did their best to weed out all other plants. Any plant that grew uninvited in cropland was called a weed and became an enemy. In many cases, the word "weed" became part of their common name: Joe Pye weed, horseweed, New York ironweed, milkweed, ragweed, pigweed, bindweed, smartweed, pokeweed, butterfly weed, hawkweed, tick weed, and fireweed are just some examples.

The situation under which such plants came to be considered weeds is understandable, but today their common names have stacked the emotional deck against them. We all have cultural permission to destroy a weed

Joe Pye weed is a productive and beautiful native plant that does not deserve to be called a weed.

Tick weed is another beautiful native.

Spreading fireweed blankets the ground of a forested landscape.

anywhere, anytime, because it's just a weed. Let's face it—we have a marketing issue with our native plants. What's ironic about our perception of native
plants as weeds is that we have no problem planting the highly invasive Asian
Paulownia tomentosa, for example, because its common name is the princess
tree. How bad could a princess tree be? The smelly and invasive Asian ailanthus
tree is known as the tree of heaven. That definitely sounds like something I
would like in my garden. I wonder if common milkweed would be welcome
in our gardens if we called it monarch's delight, or if we would plant New
York ironweed if it were renamed Manhattan splendor? Sneezeweed (which,
by the way, does not make you sneeze unless you dry its leaves, crush them to
a powder, and stuff it up your nose) could provide swaths of gold and orange

from July to October if we renamed it radiant sunset. Even plants not named weeds are often considered weeds if they are not drop-dead gorgeous or if they dare to reproduce in our gardens. How many times have we heard that native plants such as black cherry trees, the second most biologically productive tree in North America, look like weeds or act like weeds and so *are* weeds. It's no wonder that gardeners reflexively shun so many native plants.

WEEDS ARE OUR FRIENDS

The bad rap that has been bestowed upon many native plants is more than an undeserved shame: it's become an ecological disaster. These so-called weedy native plants support much of the animal diversity in North America, and our war against them in residential and commercial landscapes, along roadsides, and on the edges of croplands has been a primary cause of the decline of butterflies such as the monarch, thousands of species of native bees, and countless other insects that no one is monitoring. There is little doubt that without our native weeds, we would face ecosystem collapse. Here I provide just a few examples of how valuable these plants are to us all.

Goldenrod

Let's return to goldenrod for starters, because I didn't have time to sing all of its praises when my wife poked her head through the back door. First, I want to dispel a common misconception: goldenrod does not make you sneeze. Its pollen grains are large and sticky and do not float on the wind. The true culprit is ragweed, a nondescript plant that sheds its windborne pollen at the same time goldenrod is in bloom. The genus *Solidago*, which includes 100 described species in North America, is without a doubt one of nature's greatest gifts to animal life. The goldenrods collectively are one of the best groups of plants for our native bees because so many have specialized on its pollen, and it also provides nectar and pollen for our generalist bees. A sequence of goldenrod species can create beautiful blooms from mid-July through early October in middle latitude landscapes that meet the needs of generations of bees, while lending beautiful yellow hues to the landscape. Across the United States, *Solidago* is the top-ranked genus in terms of hosting the ecologically valuable caterpillars that feed our breeding birds and fall migrants, including 181 species of caterpillars, such as the brown-hooded owlet, the asteroid paint moth, the arcigera flower

Plants in the genus *Helenium*, common name sneezeweed, do not sound like something you would want in your garden, but, in fact, are!

moth, and the striped garden caterpillar. Goldenrod is also a cornerstone plant for meadows and prairies far and wide, is excellent in drought (I have never seen a goldenrod wilt), and produces the seed on which our winter bird residents, as well as the voles and mice that feed hawks, owls, weasels, coyotes and foxes, depend. Finally, goldenrod roots are thick and intertwined, and they not only prevent erosion, but they actively build topsoil and encourage rainwater infiltration rather than stormwater runoff. In this way, goldenrod helps replenish the water table and intercepts nutrients before they reach our streams, lakes, and rivers.

Asters

If we are talking about fall-blooming native plants, we cannot forget the asters. Taxonomically the group is a mess. The former New World genus *Aster* has been split into ten separate genera, with the genus *Symphyotrichum* housing most of our New World species. But most people still collectively call them

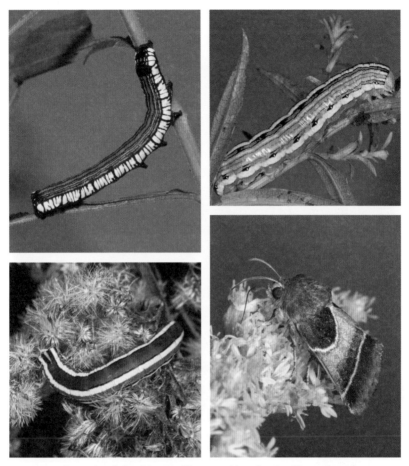

CLOCKWISE Brown-hooded owlet caterpillar, asteroid moth caterpillar, arcigera flower moth, and striped garden caterpillar

all asters. Several are showy ornamentals that few would call weeds, but many others have small white flowers and fall to the hoe all too often. Asters help fill an important niche for nectarivores such as bees and late-season butterflies because they provide nectar and pollen after most other plants have started to senesce. Indeed, asters are often the only food plants available for monarchs that have gotten a late start on their migration to Mexico. Their foliage is critical to the development of beautiful butterflies such as the pearl crescent and beautiful caterpillars such as the Asteroid moth and striped garden caterpillar. In fact, 109 species of caterpillars have been recorded using asters where I live.

Fall-blooming asters provide essential forage for migrating monarchs, even well after goldenrod blooms end.

Moreover, many bees, such as *Andrena asteroides*, will use only the pollen from asters to rear their young.

Other natives

A number of native weeds spend their days stabilizing wetlands, cleaning our waterways, limiting erosion, and providing nurseries for freshwater fish. These include pickerelweed, arrowhead, marsh bellflower, New York ironweed, gentian, swamp milkweed, New England aster, skunk cabbage, marshmallow, seedbox, Joe Pye weed, tussock sedge, meadow rue, cattail, marsh marigold,

Skunk cabbage is among dozens of native weeds that populate productive wetlands.

and hundreds of other species that don't mind getting their feet wet. Others provide the only remaining lifeline for precariously small populations of butterflies and moths, including the El Segundo blue butterfly on seacliff buckwheat, the Baltimore checkerspot butterfly on white turtlehead, the phlox moth on prairie phlox, the regal fritillary butterfly on violets, the Karner blue butterfly on wild blue lupine, and the Colorado firemoth on fireweed.

Even plants that have traits that are genuinely obnoxious to many of us are often important cogs in the biodiversity wheel of fortune. Ragweed comes to mind. I spent the better part of ages ten to thirty either sneezing my head off or lying prone with tissues stuffed up my nose because of my allergies to ragweed. Yet the ragweed genus *Ambrosia* is the eighth most productive herbaceous genus in the East, supporting caterpillar development for fifty-four species of moths, including the beautiful common spragueia and the deceptive small bird-dropping moth, while giant ragweed is the host plant for the beautiful silvery checkerspot. When you weed out plantain, you are eliminating the hitched arches moth from your yard, as well as buckeye butterflies, various tiger moths, and the giant leopard moth. One of the prettiest beetles in the world, the dogbane beetle, develops only on dogbane. The pussytoes you spray in your lawn serve as hosts for the American lady butterfly. That smartweed you despise

The endangered phlox moth, *Schinia indiana*, breeds only on prairie phlox.

Silvery checkerspot

Dogbane beetle

Smartweed caterpillar

Globemallow wingdinger

produces the lovely smartweed caterpillar. And those blackberries that drive my poor wife crazy give us the blackberry looper, the small purplish gray, the large lace-border, the spotted apatelodes, the io, the checkered-fringe prominent caterpillar, the long-winged dagger moth, the paddle caterpillar, the lunate zale, and 154 more caterpillar species—not to mention those berries that are irresistible to so many birds. If you live in the West and decide not to weed out your globemallow, you may attract *Chamaeclea pernana*, sometimes known as the globemallow wingdinger—one of the loveliest moths in the country, despite the silly common name. The point is, as we rid our landscapes and our agricultural borders of the native plants we so carelessly call weeds, we cripple our local ecosystems. I am not asking that we share our prettiest plantings with anything that wants to grow there, or that we tolerate introduced pests such as chickweed and Johnson grass in our gardens. I am suggesting that we consider giving the many native species we have labeled weeds a place of their own somewhere in our landscapes so they can help us be better stewards of local ecosystems.

CHAPTER TEN

Will It Work?

Those who dwell, as scientists or laymen, among the beauties and
mysteries of the earth are never alone or weary of life.

—RACHEL CARSON

IN 1620, WHEN THE PILGRIMS LANDED at what is now Provincetown at the
lower tip of Cape Cod, they found the potable water they had been looking
for, but they also found a near continuous span of well-spaced mature trees.
By the early 1800s, however, the entire cape had been clearcut for settlement
and sheep herding, with nary a tree remaining. Much the same story can be
told for most of New England, New York state, and Pennsylvania. Once heavily
forested, these areas were logged, first to make way for an agrarian economy,
and then massively to support the wood needed for the Union troops during
the Civil War. By 1899, Williamsport, Pennsylvania, became the logging capital
of the world, with a peak annual harvest of 2.3 billion board feet of softwood and
another 420 million board feet of hardwood (Stranahan 1993). The slash left by
industrial logging was frequently ignited by sparks from passing steam engines,
and huge areas burned repeatedly and for weeks at a time. Erosion washed the
topsoil into streams and rivers, fowling many waterways for decades, and what

was left has been described as a lifeless moonscape of burned-out stumps and gullies. By the turn of the twentieth century, vast tracks of the East were in ecological ruins.

NATURE'S RESILIENCY

The level of such destruction was shameful and unnecessary, yet it fortunately occurred before we had unleashed the ecological tumors we call invasive species. This is significant, because it meant that nature could repair herself as she always had, through secondary succession. Once the logging and burning stopped, nature arose from the ashes like a great phoenix, and the trees grew back. True, the species composition of the returning forests had changed from large swaths of hemlock and pine to forests dominated by oaks and cherries. But as the plants regenerated, so did the wildlife they supported and the ecosystem services they produced.

Though it seems miraculous—and in many ways, I suppose it is—such examples of nature's resiliency are not uncommon. From the return of life to the volcanic ruins of Krakatoa, so eloquently recounted by E. O. Wilson in *The Diversity of Life* (1992); to the rewilding of the area surrounding the Chernobyl nuclear meltdown in the Soviet Union, despite radioactive pollution; to the transformation of Ohio's Cuyahoga River from industrial sludge and fire to our most recent national park, we have witnessed time and again how quickly and completely nature can restore herself from utter destruction. Nature's inherent resiliency is the primary source of my motivation to change the landscaping paradigm from one that excludes the natural world to one that embraces it. Nature is constantly affirming that this is not a pipe dream: it will work, despite the enormous scale of change I am proposing and regardless of the ecological starting point! If left on their own, even areas from which nearly all life has been removed will eventually heal, but the healing can be unbelievably fast when we help.

As of 2007, approximately 61 million acres were classified as urban use areas, and 103 million acres were classified as rural residential use areas in the United

The Cuyahoga River in Ohio, today a beautiful national park, was once so polluted that it famously burst into flames.

In appropriate conditions, nature can repair herself remarkably quickly, as demonstrated in this woodland setting in Ohio, once the site of J. D. Rockefeller's mansion, which burned down in 1917.

States (Nickerson et al. 2011). Combined, this is nearly the size of the state of Texas and does not include the acres of landscapable infrastructure that supports us: the 424,000 acres dedicated to our eighty-two largest airports (Denver International Airport, for example, is almost two-and-a-half times bigger than Manhattan), the 15,372 golf courses occupying 2.3 million acres, the 5534 hospitals, the 5300 colleges and universities, the 1.7 million corporations and their expansive landscapes, and so on. My point is this: each of the acres we have developed for specific human goals is an opportunity to add to Homegrown National Park. We already are actively managing nearly all of our privately owned lands and much of the public spaces in the United States. We simply need to include ecological function in our management plans to keep the sixth mass extinction at bay.

CAN WE DO THIS IN CITIES?

If asked whether cities can play important roles in conservation, most people—indeed, most conservation biologists—would respond with an emphatic "No." It seems intuitively obvious that areas, often huge areas, in which cement has replaced the plants that support animal life simply could not provide the food or shelter required by species in trouble. I must confess that, until recently, I held that opinion as well. It's true that cities without productive flora will remain part of the problem rather than the solution. But in recent years, many cities have launched projects that demonstrate the real conservation potential of urban landscapes. Cement need not be the default landscape in urban environments, and when native plants are returned to cityscapes, the animals that need those plants often return as well.

On 18 June 2014, I spent a few short minutes on New York City's High Line, the long-abandoned elevated rail line in Manhattan that has been converted to a pedestrian walkway lined primarily with indigenous trees, shrubs, and perennials. The plantings are not large, maybe ten feet wide along one or both sides of the walkway that winds over car-choked streets among towering skyscrapers in the city's Chelsea neighborhood. The High Line is a popular destination with New Yorkers, and it is packed with people most days and evenings.

I had expected to see beautiful blooming plants among the woodies, but no native bees or butterflies using those blooms. I did not think that populations of pollinators could survive within—or even find—such a tiny island of habitat amidst a sprawling sea of bricks, mortar, and steel. Wrong again! Within five minutes, I saw four species of native bees foraging on butterfly weed, including two species of leafcutter bees that I had never before encountered in my travels. Had I spent more time on the High Line, I'm sure I would have seen even more bee species.

Leafcutter bees reproduce on nursery packages that they make by wrapping pollen inside leaf sections snipped from plants with thin, pliable leaves. Redbud leaves are favorite targets, and fortunately for the leafcutters, redbud is a frequent understory tree in the High Line. I'm not sure that these trees were intentionally provided for the bees, but because both the pollen and the

Here Comes the
Neighborhood

10
HUDSON YARDS

A leafcutter bee seeks nectar from butterfly weed on New York City's High Line.

Redbud leaves show the characteristic half-moon sections removed by leafcutter bees.

leaves required for leafcutters to reproduce are plentiful on the High Line, the leafcutters have successfully taken up residence here. Apparently, they are not bothered by the people walking only a few feet away or by the constant street traffic below. The bees have what they need and they appear to be thriving.

If I was surprised to see leafcutter populations, imagine how I felt when I saw three nectaring monarch butterflies a few minutes later! Those were the first monarchs I had seen anywhere in the East in 2014, and that was not for lack of looking. I had recently conducted a butterfly "bioblitz" within acres of excellent milkweed meadows at Mt. Cuba Center in Hockessin, Delaware. No monarchs. Cindy and I had been watching our own milkweed patches at home every day, where we regularly saw tiger, spicebush, and zebra swallowtails; great spangled and meadow fritillaries; sulphur and cabbage white butterflies; and many skippers. But no monarchs. Yet in Midtown Manhattan, I saw three nectaring monarchs. Whether they stayed to lay eggs on the many milkweed plantings I cannot say, because my time on the High Line was up and I had to move on. But I saw no reason why those monarchs could not have successfully reproduced along that thin strip of habitat on Manhattan Island.

Even the thin strip of plantings in the High Line in Manhattan is sufficient to support some pollinators and birds.

My trip to the High Line convinced me that the only reason our cities do not support a greater diversity of animals is that cities were not designed to do so. Most cities do not contain enough of the native plants that animal communities need to survive and reproduce; it's not that the animals cannot find the plants if they are there, or that the animals cannot reach these plants from good habitat beyond city limits. I am generalizing, of course; cities will never be havens for box turtles, salamanders, snakes, or other animals vulnerable to being squashed by zooming traffic. But there are many other animals, particularly the birds and the bees, that will be able to thrive in city environments if we simply restore the plants that support them in the landscape. Diverse native plant communities will generate diverse animal communities wherever they are

But are urban areas still capable of supporting native plants without intense gardening efforts? American botanist and author Peter Del Tredici has argued that we have transformed city environments so extensively from soil to sky that the native plant communities that once existed at those sites can no longer do so (Del Tredici 2006). Topsoil is long gone from city lots, the copious amounts of concrete in cities have dramatically raised the pH of most city soils, and we have loaded the ground with heavy metals and backfill trash, destroying the organic A horizon and complex soil biota on which so many plants depend. Moreover, many urban landscapes are highly stressful environments for plants, particularly in terms of air pollution, water availability, and heat extremes.

Nevertheless, plants do grow in cities. We see them every day along railroad beds, abandoned lots, and even sprouting from cracks in sidewalks. Often these are tough plants with great dispersal mechanisms, high reproductive potential, and superior competitive abilities—all traits that characterize our most aggressive invasive species from overseas. In fact, we are so used to seeing paulownia, ailanthus, and phragmites in city environments that we have come to believe that only plants from someplace else can do well in urban conditions. This particular urban legend may lead the list of illogical notions about native plants. Del Tredici is absolutely correct when he says that, in many cases, the original composition of plant species once found on city sites can no longer

Horticultural recommendations for growing some species, such as vine maple (*Acer circinatum*) from the Pacific Northwest, suggest regular watering, particularly during the summer; yet here, a genotype adapted to the extreme conditions of a lava flow shows that many of our native plants can fare well in less-than-perfect environments if we look for the appropriate genotypes for hot, dry conditions.

survive under current city conditions. But that is not to say it is impossible to assemble any combination of native plant species that will do well in cities. Do we have native plants that can handle a high pH, droughty conditions, air pollution, and high temperatures? Yes, we do. Can many of these plants support productive food webs, even in cityscapes? Yes, they can. Indeed, one of the primary factors that stimulated the creation of the High Line was the discovery that so many native plants had colonized the abandoned rail line with no help at all from us. Do we have a definitive list of which native plants are likely to do well in rough city conditions? Not yet; but we can gain this knowledge if we decide it's important.

It has never been a priority to look for native plants, or for genotypes of native favorites already in the horticultural trade, that would do well in cityscapes. But that doesn't mean it cannot be done. Internationally acclaimed landscape designer W. Gary Smith once suggested to me that we could seek natural genetic variants that are good at handling particular types of city stresses, even if the parent species does not have a reputation for city hardiness. For example, if we want red maple as a street tree in Philadelphia, instead of looking for a red maple with the most spectacular fall color, we could instead look for a red maple genotype that is doing well in the rocky outcrops of the Appalachian Mountains. Despite being known as a swamp trees, such red maple variants exist and have already been subjected to thousands of generations of natural selection for survival in thin soils under wide temperature swings. We could call it *Acer rubrum* 'Street Corner', and as long as we propagated it through crosses with other rocky outcrop red maples, we wouldn't need to clone it to maintain its street-hardiness trait.

THE POWER OF URBAN LOTS

The wildlife thriving on the High Line is impressive, but it doesn't hold a candle to the species that regularly use Pam Karlson's yard near the heart of Chicago. It is not paved over, but her yard is urban in every other sense of the word. She lives a half block north of Chicago's Kennedy Expressway and directly adjacent to one of the runways at O'Hare International Airport. A nearby hospital and firehouse add sirens to the roar of incoming jets. She is completely surrounded by major streets, so her property does not benefit from connectivity with preserved land. Suffice it to say that Pam's property is exposed to lots of city sounds, light pollution, cars, trucks, exhaust, and more.

Pam's yard is not large; the portion dedicated to her garden is a tenth of an acre—three times smaller than the average lot size for a new single-family home sold in the United States in 2013. A paved alley abuts her backyard garden, and her neighbors all have typical urban yards with driveways, sheds, grass, and very little garden space. Because Pam's property is small and isolated from a source of colonizing species within a sea of uninhabitable urbanscapes, Wilson's island biogeography theory predicts that there should be little wildlife in her yard (MacArthur and Wilson 1967). Fortunately, she is not familiar with the island biogeography theory. Instead, she has spent the last twenty-five years slowly and single-handedly restoring ecological function in her garden by adding and replacing plants. To date, she has added sixty species of native plants to the mix that existed when she moved in. Most of these are perennials that provide forage for native bees throughout the season, but she has also added a green hawthorn and serviceberry that keep her birds supplied with fruits and insects. A large silver maple arches over her yard with the potential to produce 287 species of caterpillars, while her neighbor's yard contributes a tall honey locust to the canopy where orioles, flycatchers, and brown creepers rest and forage. A row of northern white cedars along her fence provides a privacy screen for her, but also an excellent roosting site for her birds. The crowning hardscape of her urban oasis is an elaborate water feature that provides a reliable source of clean water for resident and migrating wildlife.

Pam Karlson's efforts have paid off beyond her wildest dreams. Not counting flyover species, she has recorded 103 species of birds in her yard, including her latest visitor, a woodcock! If I were looking for a woodcock for my bird list, I would never have looked in urban Chicago, but she has now made this find a possibility. Her favorite time of day is late afternoon happy hour in her yard, where she unwinds with binoculars, her camera, and a glass of wine.

' CAN WE RESTORE THE SUBURBS?

If restoration is possible under the most challenging urban conditions, think how easy it will be in suburbia, where much of the soil remains, air pollution is far less of a problem, and sources of colonizing plants and animals still exist in habitat patches nearby. Years ago, I lived in a northern Maryland development called Glen Farms. I was attracted to Glen Farms for one reason: its trees. Unlike all the other developments in the vicinity, the houses in this development were built without bulldozing the mature oaks, beeches, black gum, and tulip trees that were dense enough to provide nearly complete canopy cover, but not so dense that homes couldn't be built without disturbing them. Other than that, it was a typical development. All the homes had established lawns, and most homeowners had inserted their favorite Asian ornamentals as understory to the native canopy species. The first spring after I had moved in, I was visited by a scarlet tanager male who had set up a breeding territory in my front yard. I learned about this male because of his persistent and admittedly somewhat annoying habit of attacking the tanager he saw reflected in my living room window. Each morning when the light was just right, he would battle his own image for about a half hour. He never managed to scare the image away, but neither did he stop trying.

What's interesting about that scarlet tanager is not that he was eager to chase other male tanagers out of my yard, but that he was breeding in my yard at all. Scarlet tanagers are forest-dependent birds—"avoiders" in the parlance of Robert Blair (1996)—that supposedly will not breed in human-dominated landscapes. This was suburbia, not a forest interior, yet there he was, doing what all breeding tanagers do to protect their territories. Perhaps he didn't know he was supposed to breed only in large forest tracts with no human disturbance, or, more likely, he saw Glen Farms with all of its trees as a suitable forest habitat. He didn't care about the houses and lawnmowers and bicycles and dogs and roads and about a hundred families of humans. He had what he needed: large, productive trees that supplied the cover and food he required to reproduce successfully.

Since my weeks with the scarlet tanager, I have encountered several additional situations that have made me question the veracity of the avoider concept. Last spring, I watched a red-shouldered hawk raise its young off the frogs it caught in a nearby wetland that was designed to filter nitrogen-rich water. My bird book tells me I should look for red-shouldered hawks in deciduous woodlands, often near rivers and swamps. Well, there was a wetland, but the hawk's nest was in the middle of the University of Delaware campus in a tree outside of Townsend Hall, where my office is. This red-shouldered hawk bred successfully in the middle of a busy suburban environment designed to educate

Scarlet tanagers can live happily in suburbia as long as plenty of large deciduous native trees are present.

Birds thought to be avoiders, such as ovenbirds, which require understory vegetation and lots of leaf litter, may not mind humans per se but are absent from our built landscapes because they are missing a vital resource.

16,000 students. Humans and their stuff didn't seem to bother this family of hawks one bit.

I cannot tell you how many times Cindy and I have stayed in campgrounds so crowded that walking to the bathroom without being hit by a herd of young cyclers was an achievement, and so noisy that we couldn't sleep at night. Yet we have always encountered breeding interior-forest birds amidst the mayhem. I remember one campsite, where not twenty feet from our tent, I photographed ovenbirds, black-throated blue warblers, and yellow-bellied sapsuckers feeding young in nearby nests. These birds went about their business as if we humans were the wood bison or elk they would have encountered in those same forests centuries earlier. Despite the noise and confusion, they were there because the habitat was rich with the food and shelter they needed.

These stories, though true enough, are considered anecdotal evidence by scientists. I hate to think of my beautiful scarlet tanager as a mere anecdote, but science says we need to test hypotheses before we accept them. Fair enough, for that is what science is—a discipline that tests hypotheses. In this context, we need to design experiments to test whether birds heretofore considered species that avoid humans may actually be absent from our neighborhoods because of the loss of some vital habitat component instead of the presence of humans. These species require tree snags with nest holes, canopy shade for temperature regulation and/or cover, appropriate amounts of insect food sources, leaf litter for ground foragers, and more. Would they take up residence in neighborhoods that contained what they needed to survive and reproduce? My scarlet tanager and campground experiences suggest that they would, but it needs to be formally tested to know for sure.

SUBURBAN CHALLENGES

Clearly the low-hanging fruit in this endeavor is the land surrounding our homes—the land we own and can easily manipulate ourselves, starting today. Anyone who has sought certification for a wildlife-friendly habitat from the National Wildlife Federation, Audubon, or many smaller groups knows the requirements. You must design a yard that supplies water, food, and shelter. This seems simple enough, but there are a few sticking points that give many would-be conservationists pause. Is my yard really large enough to make a difference? How do I square my new landscape with the unforgiving dictates of my homeowner association? If I plant wildlife-friendly trees such as oaks, won't they eventually fall on my house or car? All these good questions deserve good answers.

First, let's talk about yard size. Many people believe that if they don't own acres and acres, their yard is simply too small to build a forest, a savannah, or a prairie large enough to support much wildlife. That might be true if your property were an isolated island in the middle of the ocean. But it's not. Your property abuts your neighbor's property, which abuts another property, and so on. It is more accurate to envision your property as one small piece of a giant puzzle, which, when assembled, has the potential to form a beautiful ecological picture. If the target habitat in your area is an oak savannah, then all you need to do is create a small section of that savannah. On very small properties, that may be only one oak tree, a blueberry bush beneath it, and some spring ephemerals beneath that. If your neighbors do the same, the greater habitat will come into being. And if neighbors refuse to add productive plants to their yards and just stick with the barren lawn designs that now dominate our landscapes, your property will still provide some of the food that more mobile species require, even if it doesn't meet all of their needs throughout their lives.

This is particularly true for migrating birds. Migrants fly right through our cities and suburbs; they do not attempt to avoid them. Good thing, because that would be impossible these days. Migrating birds fly all night and then rest and refuel during the day. By refuel, I mean they eat insects. Most migrants that fly all night have used up their energy reserves by morning light. To continue our analogy, they are out of gas and won't be able to resume their migration until they gas up again. This is not a trivial task; in one day at a stopover site, a migrating bird will typically eat enough insects to increase its body weight 30 to 50 percent (Faaborg 2002). If it ends its night flight around 4:30 a.m. and happens to come down in a city or suburb, where the only trees are Asian Callery pears, ginkgos, and zelkovas—a typical mix of street trees these days— the gas pump will be dry and there will be no insects to eat. This is where the homeowner with a small yard can make the difference between continued life and death by starvation for challenged migrants. By planting a keystone tree such as an oak, a native *Prunus* species, a willow, a cottonwood, some blueberry bushes, a birch, or a hickory, homeowners are throwing a lifeline to migrants that will enable them to move through heavily built areas without becoming too weak to continue their flight.

But, you may ask, if I plant that oak tree in my yard, won't it eventually fall on my house? We certainly hear about such mishaps after every derecho, bomb cyclone, and hurricane that comes along, and sometimes after a really big rain. Though small, the risk of large trees falling on our possessions—or worse, on us—is real enough. But do we have to risk life and limb to have a living land-scape? Of course not. Life is not risk-free, but treefalls are something we can and should manage. The obvious solution most people would offer is to plant all trees with the potential to become large far from the house or driveway. This works if you own lots of land, but it's not an option on small properties. Not planting large trees also deprives your house of the cooling that shade from trees provides in the summer and the windbreak trees provide during winter.

Fortunately, there is another solution to this vexing problem and it comes from the way trees grow in nature. Trees evolved to grow together in a forest. They intertwine their roots, forming a root matrix that is nearly impossible to uproot. Forest trees with interlocked roots may snap off in big winds, but they typically don't uproot. Because aesthetics have trumped function for so long, we have planted large, isolated specimen trees ready to blow over nearly everywhere. If we change our goal from creating majestic specimen trees to picturesque groves of trees, the interlocking effect of root matrices will be

This eroded stream edge clearly shows how tree roots in nature interlock with one another, making tree groves resilient to high winds.

strongest. Just down the street from me stands a pair of large white oaks whose trunks are no more than three feet from each other. Even though the road that was built next to them many decades ago severed half of their roots, they have withstood every storm since without toppling. Few arborists would suggest planting trees on a three-foot center, but if we planted our trees in groups of three or more on ten-foot centers, the resulting root matrix would keep them locked in place through thick and thin. None of the trees would develop into a single majestic specimen tree, but together they would form a single grove of trees that the eye will take in just as if they were one large tree. Planting tree groves will also protect against the domino effect. Every time we take down a tree, we make the remaining trees more vulnerable to straight-line winds. There is one catch to this approach, however: the trees must be planted young, so their roots can interlock as they grow. Transplanting five-inch caliper trees that are twenty feet tall for instant gratification is a poor way to achieve interlocking roots with any strength.

OK, we now have our trees and we understand the value of even small restorations. But will the homeowner association allow us to deviate from the status quo? Of all the stumbling blocks to building Homegrown National

A grove of 100-foot Douglas firs in a park in Portland, Oregon, shows how closely trees can and do grow together with no ill effects.

One or two mower widths of manicured lawn provide an ideal cue for care, demonstrating that the associated landscape is intentional and not a product of neglect.

Park, this may be the biggest. But contrary to popular belief, HOAs are not insurmountable obstacles for two reasons. First of all, HOA rules were established by people, not by Martians. Because we are also people, we can change those rules. Most HOAs were established decades ago to dictate a culture that would protect property values. The goal was to keep the neighborhood neat and high class; broken-down cars and rusted refrigerators in the front lawn were frowned upon. So were landscapes that looked untended and weedy. The message these rules were designed to convey was that everyone who lived within was a good citizen with upper-class values. The visual advertisement of those values included proper land stewardship, which marketing folks had taught us was impeccable lawns dotted with a few Asian ornamentals.

And there we have the argument that can bring HOAs into the twenty- first century. For today, we cannot be good citizens with responsible, community-minded values if we keep our precious land dead as a rock. We now recognize that the message outdated HOAs send their neighbors is destructively self-centered: "We don't care how our rules affect your ecosystem. Your watershed, climate, and pollinators be damned!" Our task, then, is to rewrite HOA guidelines. We must join our HOA committees and educate the other members about the many ecological roles our landscapes must now play, how every landscape must be designed to support diverse food webs that contain both

This pink Florida dogwood illustrates how understory trees can create vertical layers of habitat in managed landscapes.

herbivores and their natural enemies, and how our yards must support both generalist and specialist pollinators as well as manage our watersheds and store as much carbon as possible in their plants and soils.

Few HOAS restrict particular plant species or the abundance of plants we have in our yards. Restrictions are more often applied to how and where we use these plants in our landscape designs. Many people equate native landscaping with a total lack of planning, where the property is just left to go wild. We can combat this misconception by designing artful landscapes that will differ from traditional landscape designs in only three ways: they will have less lawn, with more plants in total, and more of those plants will be the powerhouse species that drive food webs and support pollinators.

The single required feature of an HOA-approved landscape is that it be cared for. It's easy to put cues for care front and center in our yard for all to see. A yard within Homegrown National Park will have less lawn than it used to, but the lawn that is retained will be manicured nonetheless. Nothing advertises your commitment to neighborhood standards like a manicured lawn. Our lonely specimen trees will now become the tallest members of a layered landscape, arching over understory species such as Florida dogwood, witch hazel, and silverbells; shrubs such as various native viburnums and hazelnuts; and groundcovers of violets, mayapples, and native pachysandra. But these designed plant communities can be formalized with a neatly trimmed grass border that clearly defines their intentional nature.

One of the biggest challenges for suburban dwellers is to create plantings that deliver the benefits of a meadow or prairie in areas on full display to the public. This is a particularly important issue in the Midwest, where prairie ecosystems are most sustainable. The perennials included in such plantings look great when they are in bloom, and if carefully selected plant species are used, they can be in bloom throughout most of the growing season. But what about late fall, winter, and early spring, when last year's plants look like a mass of dead stalks? Many people are tempted to neaten up in the fall by mowing their meadows, but that destroys the native bees that are overwintering in those dead stalks as well as the eggs of meadow katydids and praying mantids, the chrysalises of swallowtails, and the cocoons of countless moths. Fall mowing also eliminates the seeds that goldfinches, sparrows, and juncos rely on all winter long.

One solution proposed by the Little Garden Club of Wilmette, a progressive club in Illinois, again uses lawn beauty strips to define what they call pocket

An example of a diverse pocket prairie tastefully planted in a front yard.

prairies. Pocket prairies can be located almost anywhere and shaped in attractive contours that take advantage of unused spaces in your yard. They can be as small as three-by-seven feet and still provide pollen and nectar for flower visitors as well as nectar and host plants for monarchs. Appropriate species selection can produce blooming pocket prairies in either sun or shade. Winter interest can be added with native bunch grasses such as wild oats (*Chasmanthium latifolium*), prairie dropseed (*Sporobolus heterolepsis*), and little bluestem (*Schizachyrium scoparium*) that turn rich browns and grays in dormancy. Heather Holm, an expert on Midwest native bees, has discovered that most bee species that spend the winter in the stalks of senescent prairie forbs do so within one foot of the ground. If she is right, we could cut nearly three-fourths of taller stalks off without reducing bee survival.

THE DRAW OF A BUBBLER

In the early 1980s, Margy Terpstra found a beautiful bird in her yard—beautiful, but dead. It was a Kentucky warbler killed by a window strike. She knew very little about birds, but the tiny body in her hands triggered a flood of questions and a keen desire to find the answers. What could it be? Where did it come from? How many others like it were out there? Her husband, Dan, encouraged her nascent interest in birds with a new pair of binoculars and a birding class at the St. Louis Zoo. Soon Margy was going on bird walks and keeping a running list of the bird species she had seen in her suburban St. Louis yard.

In 1996, after two kids had come and gone, Margy and Dan made plans for their next stage of life. First, they decided to move. They wanted more land and less house, and like many homeowners, they had a choice: a traditional large lawn dotted with a few small Asian trees or a smaller lawn with several mature native trees. Margy's growing passion for birds made this choice an easy one. They found the perfect corner lot with several large trees in Kirkwood, Missouri. The lot was about two-thirds of an acre in size—not tiny, but not huge—and included a diversity of deciduous trees, including oaks, hickories, cherries, ashes, and maples. But the understory was choked with invasive woodies, particularly bush honeysuckle. Once again Dan was supportive, and soon he had removed the honeysuckle, creating room for native blackhaw viburnum, mayapple, Virginia creeper, roughleaf dogwood, and gray dogwood.

Margy had learned that birds need food, shelter, and water during migration and while breeding. The native plants in her yard were providing food and cover, but where would birds find a convenient sheltered source of clean water? Inspired by a design she saw at nearby Tower Grove Park, Margy and Dan created a bubbler: a small, continuously circulating water source that splashed gently into a shallow, pebble-lined pool. Margy discovered that clean, shallow water combined with splashing sounds are irresistible to most birds year-round, but particularly during migration when Neotropical migrants are tired, thirsty, and in need of a bath. They built the bubbler within view of their back porch and immediately discovered what a birder's (and bird photographer's) paradise they had created. Warblers by the dozens stopped by regularly for a drink and a splash. To date, the Terpstras have seen 35 species of warblers (plus one hybrid), including the exquisite blackburnian, magnolia, and bay-breasted warblers, and 120 total bird species at their bubbler. The total species count for their yard is 149, including every warbler species that occurs in the state of Missouri. In fact, some Missouri state records were made by Margy herself at her bubbler, and every year she sees new visitors. Bubbler visitors were so beautiful and so close that Margy couldn't

Margy and Dan Terpstra's bubbler, a warbler magnet in Kirkwood, Missouri.

resist the urge to take their pictures. And so, again with Dan's help, she has become a top-notch bird photographer and videographer. Her images are so good that she and Dan now host a weekly blog featuring the birds and other wildlife that visit or breed in their yard.

Did the Terpstras need special training to help so many birds (and have so much fun)? Did they need a degree in conservation biology, ornithology, or ecology to succeed? Did they need to forsake their normal routines of work and socialization and move to a wilderness research station with steel-spring bunk beds? Of course not! Margy spent twenty-one years as a dental hygienist and Dan is a mechanical engineer. What they did have was the desire to turn their tiny corner of the world into a haven for the life around them, which has become an enormously rewarding experience for them and a mechanism to connect with likeminded people. By shifting their parental instincts from their own kids toward ecosystem stewardship, the Terpstras are meeting the needs of migrants that must fly thousands of miles through human-dominated landscapes twice a year. This makes rest stops such as the Terpstras' yard essential for migrants, especially in large cities. And each stop, no matter its size, has ecological value. Margy and Dan have shown us all how effective one couple can be in their spare time. I cannot think of better role models.

IT IS ALREADY HAPPENING

As Pam Karlson, Margy and Dan Terpstra, and the Little Garden Club of Wilmette have shown us, we can do this, and others are doing it all over the country. And the more we do it—the more our culture accepts that humans will not survive the destruction of nature—the easier it will be to convince homeowners to restore their properties. Pam, Margy, Dan, and hundreds of other individuals across the country are pioneers, founding mothers and fathers of Homegrown National Park, but some large organizations have also joined the effort. One of many examples of rapidly growing initiatives across the country is the St. Louis Audubon Society's Bring Conservation Home program that was introduced to the region in March 2012. For a small fee, urban landowners receive an onsite habitat consultation from a team of trained advisors who provide recommendations on how to improve the landscape for the benefit of native plants, animals, people, and the local ecosystem. The resulting report covers the removal of invasive plants, the installation of a native plant landscape, stormwater conservation practices, pollinator gardening, and other wildlife stewardship practices. As of 2017, 860 homeowners had enrolled in the program, resulting in the ecological improvement of more than 400 suburban acres. In fact, the program has been so successful that habitat advisors are having trouble keeping up with the demand.

Of course, the concept of a homeowner-driven certified habitat program is not at all new; it was the brainchild of the National Wildlife Federation way back in the early 1970s. For forty-five years, the NWF Certified Wildlife Habitat program has provided a model for dozens of smaller programs across the nation and has grown from just 249 certified habitats in 1973 to 217,000 in 2018, impacting more than 2.5 million acres. The program has also recognized the restoration potential of land well beyond our yards and now includes certified college campuses, schoolyards, corporate landscapes, botanic gardens, zoos, and indeed more than 100 entire communities. These certified habitats are the first pieces of the Homegrown National Park puzzle, the edges if you will, but crucial center pieces are still needed to complete the picture of a sustainable relationship between humans and nature. What is interesting and encouraging to me is that all of this has occurred despite our society's growing disconnect from the natural world. Think how quickly the entire country will become engaged when our need for robust and diverse ecosystems in all of our human-dominated landscapes becomes common knowledge!

CHAPTER ELEVEN

What Each of Us Can Do

There is in fact no distinction between the fate of the land and the fate
of the people. When one is abused, the other suffers.

—WENDELL BERRY

THERE IS HARDLY a page in this book that doesn't imply or directly describe
something each of us can do to contribute to Homegrown National Park. But
no more beating around the bush—in this chapter, I briefly outline ten concrete
steps each of us can take to make Homegrown National Park a successful reality.

SHRINK THE LAWN

Most suburban, rural, and corporate landscapes have more space dedicated to
lawn than anything else. Every square foot dedicated to lawn is a square foot
that is degrading local ecosystems. Yet turfgrass is the perfect plant to walk on,
because it can take light to moderate foot traffic without dying. We have also
seen how turf can be an effective cue for care, to indicate that you are consci-
entiously caring for your landscape. A general rule of thumb, then, might be
to reduce your lawn by half. Restrict turf to wide paths that guide pedestrians
through your landscape, that draw the eye to a featured aspect of your design,

or that define beds, tree groves, or various hardscapes as being purposeful and cared for. In their 2015 book, *Planting in a Post-Wild World*, Thomas Rainer and Claudia West suggest that we think of lawn as an area rug, not wall-to-wall carpeting. Superb advice!

REMOVE INVASIVE SPECIES

When we choose plants for our landscapes, there is room for compromise, but not when it comes to introduced plants that have a history of spreading to our natural areas at the expense of native plant communities. This is a no-brainer when we think of what invasive plants do: they are ecological tumors that spread unchecked into our local ecosystems, castrating the ecosystem's ability to function. Over time, we can get control over this problem, but not until we stop planting such species as ornamentals and start investing the time and energy required to remove those we have already planted. For most homeowners, this can be accomplished without too much blood, sweat, and tears. For folks who own larger properties, it can be a substantial challenge. Nevertheless, if every property owner removed the invasive plants from his or her land, the goal of ridding the United States of these troublemakers, or at least reducing their seed rain to manageable levels, would be largely realized. We could then focus on our public lands at our leisure.

PLANT KEYSTONE GENERA

Our research at the University of Delaware has shown that a few genera of native plants, or keystone genera, form the backbone of local ecosystems, particularly in terms of producing the food that fuels insects. Landscapes that do not contain one or more species from keystone genera will have failed food webs, even if the diversity of other plants is very high. Throughout most of the United States, native oaks, cherries, willows, birches, cottonwoods, and elms are the top woody producers, while goldenrods, asters, and sunflowers lead the herbaceous pack, but you can find a list of both woody and herbaceous plant genera that are best at supporting local food webs in your county at the National Wildlife Federation's Native Plant Finder website (https://www.nwf.org/NativePlantFinder). Add these powerhouses to your landscape, and you will be well on your way to contributing to Homegrown National Park.

BE GENEROUS WITH YOUR PLANTINGS

To realize the ecological potential of our landscapes, most of us have to increase the abundance and diversity of our plantings. If you have one tree in your yard, consider adding two more. Remember that you are planting groves of trees at the same density at which they occur naturally in a forest. They may seem crowded at first, but they will interlock their roots and support one another in high winds. They will also supply the cover many animals need to feel comfortable near humans. Be sure to add vertical heterogeneity to your plantings by adding understory trees and shrubs to your yard. Don't fret too much about your plant choice decisions; your plant choices are not finalities; they are part of a process. A wise friend of mine suggests we interview our plants before we hire them to do a particular job in our landscape. In other words, we should plant one or two individuals and see how they do—a probationary period, if you will—before we invest in appropriate numbers of the species in question.

PLANT FOR SPECIALIST POLLINATORS

Because so many of our native bee species specialize on particular plant groups when gathering pollen for their larvae, it is essential that we meet their needs in our pollinator gardens. The common generalist honey bees and bumblebee species will use plants needed by specialists as well, so by planting for the specialists, we have planted for all bee species. We still have much to learn about which plant genera support the most specialists, but we already know some of the best plants for specialists in most parts of the country. These include perennial sunflowers (*Helianthus* spp.), various goldenrods (*Solidago* spp.), native willows (*Salix* spp.), asters (*Symphyotrichum* spp.), and blueberries (*Vaccinium* spp.). Including these plants in our gardens, along with the greatest diversity of native flowering plants we can muster, is our best defense against losing local native bee species.

NETWORK WITH NEIGHBORS

In most cases, we can increase the chances that a particular conservation effort will succeed as the area being conserved increases. The best way to have a bigger impact in suburban and urban landscapes is to team up with likeminded neighbors to focus on one or more conservation goals. Let's say you want to help the monarch butterfly. You may own only a quarter-acre lot with space for just one small milkweed patch, but if you join forces with seven other small property owners in your neighborhood, combined you will have two acres in which to plant many milkweed patches as well as pocket prairies with fall asters and goldenrods that will supply the nectar for migrating monarchs. Using a similar approach can help save our declining native bees, many hummingbirds, and, who knows, maybe even endangered species such as the Karner blue butterfly or beautiful prairie phlox moth. Social media can make it easy to find neighbors who are willing to join your conservation community. Websites and apps such as Nextdoor and YardMap enable you to connect and coordinate with others in your neighborhood more easily than ever before.

BUILD A CONSERVATION HARDSCAPE

We can reduce the carnage we humans regularly inflict on local wildlife in ways that have nothing to do with gardening.

- Each year millions of toads, frogs, voles, and other small creatures become trapped in our window wells, where they slowly starve to death. Installing cheap window well covers can reduce these needless deaths to zero.

Installing cheap covers can prevent small creatures from entering and becoming trapped in a window well.

- Use motion sensor security lights that light up only when an intruder enters your yard. Blazing security lights are ecological traps that kill thousands of moths, increase your carbon footprint, and do not improve security over lights with motion sensors.

- Set your mower height no lower than three inches (four inches is better). Not only will you have healthier, greener grass that requires less watering, but there is a good chance you will be able to mow right over a box turtle without killing it. And try not to mow in the evening. As dusk approaches, many nocturnal species leave their hiding places and are vulnerable to being pulverized by your mower.

- Install a bubbler. Small water features with gentle gurgling sounds don't take much space but are irresistible to migrating and resident birds.

- Rather than installing one large bee hotel, build several small ones with only four or five holes each. Then disperse these smaller units throughout your yard. This will make it more difficult for bee predators, parasites, and diseases to be able to attack all of your native bees in one convenient place.

CREATE CATERPILLAR PUPATION SITES UNDER YOUR TREES

More than 90 percent of the caterpillars that develop on plants do not pupate on their host plants. Instead, they drop to the ground and pupate within the duff on the ground or within chambers they form underground. When we replace soil rich in organic matter with lawn compacted by frequent mowing, it is difficult for caterpillars to burrow into the soil or find leaf litter for their cocoons. Replace the lawn under trees with well-planted beds replete with groundcovers appropriate to your area. Large decorative rocks also provide pupation sites—but, better yet, add a fallen log or old tree stump to the bed. Many caterpillar species bore into and pupate in decaying wood, a resource not available in most yards. Finally, treasure your leaf litter. Many leaves that fall each autumn harbor small caterpillars within curled leaf margins, and dozens of caterpillar species eat fallen leaves. Replace store-bought mulch with natural leaf litter wherever you can, and if you have more leaves each fall than your beds can accommodate, that's a good sign that your beds aren't large enough.

DO NOT SPRAY OR FERTILIZE

It is self-evident that using insecticides and herbicides, particularly to produce a perfect lawn, is antithetical to the goals of Homegrown National Park. It is far less evident that fertilizers are also unnecessary. After all, if the commercials tell us to use fertilizers, we should use them, shouldn't we? Keep in mind that most North American native plants are adapted to the low nitrogen soils they encountered after the last glaciation and do not require high doses of synthetic fertilizers. In fact, highly fertilized soils favor many nitrogen-loving, invasive, non-native plant species. Creating soils rich in organic matter is entirely sufficient for healthy plants. What's more, before they are taken up by plants, most of the fertilizers we use are washed into our waterways, where they cause deadly algal blooms, red tides, and other problems.

Oppose mindless mosquito spraying by your township or HOA. Contrary to what the fogger operator may have told you, the pyrethroid-based insecticides used by mosquito foggers indiscriminately kill all insects, not just mosquitos. Ironically, targeting adult mosquitos is the worst and by far the most expensive approach to mosquito control, because mosquitos are best controlled in the larval stage. Put a five-gallon bucket of water in a sunny place in your yard and add a handful of hay or straw. After a few days, the resulting brew is irresistible to gravid (egg-filled) female mosquitos. After the mosquitos have laid their eggs, add a commercially available mosquito dunk tablet that contains *Bacillus thuringiensis* (Bt), a natural larvicide, to your bucket. The eggs will hatch and the larvae will die. This way, you control mosquitos, and only mosquitos, without the use of harmful insecticides.

EDUCATE YOUR NEIGHBORHOOD CIVIC ASSOCIATION

Many homeowners believe they cannot use more native plants in their landscape because of rules developed and enforced by their township, civic association, or homeowner association. These rules were developed back when we thought humans could exist independent of the natural world, we thought there was plenty of nature out there, and we thought that plants were just decorations and nature was nice but optional. Thankfully, we are finally recognizing that none of those notions is correct, so it is time to update the antiquated, destructive landscape regulations that were based on them. Join your HOA, educate the uninformed, lobby your township to increase the number of productive plants in public spaces, and help rewrite the rules. Change will come more quickly if we recognize that these are not black-and-white issues and there is room for compromise.

Concluding Remarks

WHEN I HAD MY DRIVING PERMIT some fifty years ago, my father and I would occasionally use the morning trip to my high school as an opportunity to practice my road skills. One day, while we walked through our garage to the car in the driveway, I repeatedly tossed the car keys in the air; I must have felt a need to demonstrate how good I was at catching them. As soon as we reached the driveway, I gave the keys a mighty toss—and they didn't come down again. I had tossed them on the roof of the house! My father, accustomed as he was to my shenanigans, nevertheless was beside himself. A dash for the ladder, a risky reach with a garden rake, and I had the keys in hand once again. I made it to school on time and my father finally relaxed, but my foolishness had put me at risk of being late for school and my father late for work, two deadlines not to be trifled with in those days.

I mention this embarrassing (but not uncommon) segment of my youth because there are parallels with our current plight. We have foolishly thrown our ecological keys on the roof. We know how to get them down, but we are running out of time and must act quickly. Nature has proven to be more resilient, more malleable, and more forgiving than I ever thought she could be, but, like my father, she is growing increasingly impatient. I can only hope that, like my father, she is willing to give us all one more chance.

Building Homegrown National Park will be the most ambitious restoration initiative ever undertaken. We will create a sustainable balance between humans and other earthlings in the United States (hey, why not the whole globe?), and we will do it by living with nature instead of living apart from it. Instead of denaturing our environment as if it didn't matter, our new national pastime will be to renature our surrounds. Restoring ecosystem function can become a goal that unites rather than divides us; Homegrown National Park will have no political, ideological, religious, cultural, or geographic boundaries because everyone—every human being on this planet—needs diverse, highly productive ecosystems to survive. We must replace our current "humans or nature" mentality with a new "humans and nature" ethic. Many of us have already begun this worthy task, and the results are immediate, encouraging, and enormously satisfying. Imagine the sense of accomplishment that will come from having a role in life's salvation.

Extinctions inevitably occur when the world changes faster than species can adapt. Humans have been changing the world ever since we became humans—but now, with our numbers and technologies, these changes easily overwhelm the ability of most organisms to adapt. For species that are specialized within the environment in which they evolved (which is nearly all of them), this is a deadly problem, but if we can destroy habitat with blinding speed, we also have the intelligence, knowledge, and ability to restore it. It remains to be seen whether we have the collective wisdom to do so, but I, for one, believe we do!

FREQUENTLY ASKED QUESTIONS

Q: **Won't climate change make restoring natives to our landscapes nearly impossible?**

A: Climate change is certainly exacerbating the many pressures we humans have already put on plant and animal communities, and it won't make restoration efforts any easier, but we must not use climate change as an excuse to do nothing. Most species of plants and animals are far more resilient to climate variability than we give them credit for. Besides, increasing the number and biomass of the plantings in our yards and public spaces is one of our most accessible and convenient tools to fight climate change. Every plant you add to your yard is built from carbon it has pulled out of the atmosphere. It also pumps carbon into your soils via its roots throughout its lifetime, improving our carbon imbalance daily.

Q: **I have heard that invasive plants disrupt ecosystems. Are ecosystems really so fragile that a single invasive plant can harm it?**

A: Can a single tumor disrupt your internal ecosystem? Indeed, it can. By its very nature, it doesn't stay a single tumor; it spreads. I like this analogy because, by definition, invasive species spread, displacing native plant and animal communities wherever they go. And they keep spreading until checked by us or by climate. Moreover, our ecosystems don't have to contend with a single invasive plant species; there are now more than 3300 species of invasive plants in North America.

Q: Why should I care if birds are disappearing? I don't even like birds.

A: You should care because birds are excellent ecological indicators—canaries in the coal mine, if you will. They would not be disappearing if the ecosystems that support them were functioning properly. And you should care whether the ecosystems that support birds are healthy because those same ecosystems support you. The species in an ecosystem are the engines that run that ecosystem. Every time we lose a species, either literally or functionally, whether it is a bird, a bee, a lizard, a plant, or a nematode too small to see, that ecosystem functions below its capacity and our life support systems are weakened.

Q: A man recently asked me where the grasshoppers of his youth had gone: As a youngster, walking through fields or even vacant lots in Illinois, or western New York state, New Jersey, and elsewhere in the summer, it seemed there was an incredible number of grasshoppers jumping around and hitting my legs as I walked. Sadly, there seem to be few, if any today. In the evenings, the sounds of crickets were everywhere. No longer. No one has explained to me why grasshoppers and crickets have disappeared.

A: Whether or not grasshoppers, crickets, and other orthopterans are disappearing depends on where you are. Fortunately, grasshoppers and crickets are doing well in many places. But, as you note, they are nearly gone in others. Anyone who uses a commercial lawn maintenance service provider will not have crickets or grasshoppers in their yard. Anyone who uses fertilizer and mows religiously will have very few grasshoppers and crickets. Any farm field growing glyphosate-ready corn or soybeans will not have the plants needed to support grasshoppers and crickets. Millions of acres that are now lawn in the United States once supported the native herbaceous plants that fed lots of grasshoppers and crickets. Grasshoppers, despite their name, depend primarily on broadleaved forbs, while crickets mostly develop on dead plant material. In pursuit of our obsession for neat landscapes, we have eliminated both in too many places. Finally, areas overrun with invasive groundcovers such as Japanese stiltgrass, vinca, or English ivy wouldn't support grasshoppers because the plants grasshoppers depend on have been replaced by species they cannot eat. We can bring grasshoppers and other insects back if we plant more of our private and public spaces with the native plant species they require.

Q: I live in an area with lots of Lyme disease. Several websites suggest keeping large lawns because they are not attractive to ticks. The sites also say I should get rid of brush piles because ticks love them. How can I contribute to Homegrown National Park without getting Lyme disease?

A: Whenever I hear this question, two platitudes immediately pop into my head: Life is not risk free, and life is a trade-off. It's true that a lawn will not support a large tick population (or anything else), and pavement supports even fewer ticks. To create a world with no ticks, we could turn everything into lawn or pavement. The risk from Lyme disease would drop to zero, but so would the probability that we will persist on this planet much longer. Let's think about what deer ticks need to complete their lifecycle, and then think about the easiest way to disrupt that lifecycle. Deer ticks do not eat native plants, leaf litter, or brush piles. Ticks stay in brushy areas because they need high humidity. Deer ticks do eat mammals (or at least their blood), and two of their favorites are white-footed mice and white-tailed deer. Although Lyme disease has always been around in very low frequencies, it reared its ugly head in the 1970s because white-tailed deer changed in abundance, from the few in the 1950s and 1960s to several times over their carrying capacity in recent years. Because too many deer are bad news for the environment, reducing deer numbers might seem like the best place to interrupt the Lyme disease cycle. We need to agree collectively as a society that having too many deer is not OK, and that it's not something we have to tolerate. I realize a single homeowner cannot bring deer populations down to ecologically safe levels, but each one of us can stop opposing deer culls at our local township meetings.

What can we do in the meantime to minimize our exposure to deer ticks? Ticks do not run after us when we go into our yards. They climb up on vegetation and quest—that is, they wait for us to walk by and then grab on when we do. So, one easy solution is to reduce your lawn to wide mowed paths, and then stay on those paths during periods of high tick infectivity (that's May and June in Southeast Pennsylvania, for example). Staying out of the woods is not an option I choose, so I remain vigilant. With a little help from my wife, I check myself after I've been playing outside. Deer ticks like bare patches of skin near waist and sock bands or tight undies, and with close inspection they can be easy to find. They also like to get between my toes. Fortunately, they avoid our hairy heads. When I find an embedded tick, I pull it off (sometimes I need tweezers for those tiny nymphs) and put antibiotic ointment on the bite site. A Lyme researcher told me years ago that the ointment kills the *Borrelia* spirochete before it gets into the bloodstream. I don't know if that is true, but I do know that I have never been infected with Lyme disease when I follow that rule (and I have been infected five times when I didn't follow it). This might seem like more aggravation than it's worth, but the joys I get from interacting with nature far outweigh the nuisance of tick checks.

Another thing we can do is landscape in a way that encourages higher mammal diversity in our neighborhoods. Research has shown that areas with foxes, chipmunks, squirrels, raccoons, groundhogs, and possums have much lower rates of Lyme disease, not because they have fewer ticks, but because those mammals are dead-end hosts for Lyme disease. Simplified landscapes that harbor only white-footed mice and deer have much higher rates of infectivity.

Q: **What is the difference between a naturalized landscape and a native landscape?**

A: Naturalized can be synonymous with invasive, or it can simply mean plants that require little to no care after they are established, like a massed planting of daffodils. In either case, the word "naturalized" refers to plants that are not members of the native plant community into which they are planted or into which they have moved on their own.

Q: **How can we hope to build Homegrown National Park in a nation so culturally and politically polarized that we cannot even agree on whether the sky is blue?**

A: The fact that we have allowed certain special interest groups to politicize the health of our environment has always baffled me. There is no one on the planet who does not require a healthy environment that includes the complex ecosystems that create our life support systems. Yet we tolerate or even vote for people who foul our nest (our only nest) for short-term profit. Fortunately, we don't need an act of Congress to restore ecological function to our landscapes. In fact, if we are just a wee bit clever, we don't need anyone's permission. We can add productive plants to our landscapes without others even realizing it or objecting to it.

Q: Plants and animals have always moved around the planet, so the arrival of new species at our shores is a natural process. So what's the big deal? If the new species are more fit than the species already here, don't the new species deserve to replace the old species?

A: I hear the "they deserve to succeed" argument quite a bit. It is true that species have always moved around the planet, but their rate of movement was extraordinarily slow compared to the rate at which we are moving plants and animals around the globe today. This matters. If new species arrive at our shores only occasionally, say once every 1000 years, they would never be numerous enough to overpower the complex resident communities, and their addition to existing communities would be so slow that local native species could adapt to their presence. Four conditions make the ecological contest between resident species and novel species inherently unfair, if we can anthropomorphize just a bit. First, new species typically arrive in novel habitats today through the actions of humans. And they usually are not imported once, but repeatedly. Think of ornamental plants from Asia. They were brought to this country and have been sold by the millions over wide geographic areas for decades. The scale of such an influx of new species into our ecosystems in no way resembles any type of dispersal that happened through natural mechanisms in the past. Second, we are importing thousands of new species all at the same moment in ecological time, so in any given place today, our native flora must simultaneously compete against dozens of introduced species. Third, in nearly every case, we humans have disturbed native plant communities with our backhoes and bulldozers before introduced species were able to establish successfully. Most introduced species would never have been able to outcompete native plant communities if those communities had not first been gutted by human development. Finally, new species are typically introduced without the predators, parasites, and diseases that keep them in check in their homelands, so they are healthier when they enter into competition against native plants that must survive attack from multiple species of herbivores and diseases. To call introduced plants more fit and thus more deserving of a place on this continent than our native plants seems a stretch when we have so unevenly stacked the playing field against our natives.

Q: **Are connected habitat fragments more vulnerable to invasion by non-native plants, and if so, should I promote isolated habitats?**

A: Connected habitats are always more sustainable than isolated ones, but skinny connections are subject to plant invasions because they are all effectively edge habitat. With this in mind, building the widest connections possible is always the best strategy. Keep in mind that well-planted neighborhoods with few invasives can serve as good biological corridors connecting natural areas for many species. Dense plantings and high plant diversity within corridors will help repel biotic invasions. But there is no getting away from heavy management. Because the corridors we are talking about will be in your own yard, you can minimize the invasion of debilitating aliens with regular vigilance.

Q: **A man recently wrote to the Virginia Native Plant Society about a conundrum concerning the healthful food qualities of the nonnative autumn olive, *Elaeagnus umbellata*: This is one of those confusing issues. Autumn olive is touted by natural food experts as being a nutritious source of the cancer fighter lycopene (eighteen times as much lycopene as tomatoes). But environmentalists and most conservationist organizations (including most state and federal agricultural agencies) have declared it an invasive species and want to destroy it. Shouldn't we value this plant for what its berries could give us?**

A: Every plant can be evaluated through a cost-benefit analysis. The ecological costs of autumn olive are enormous. They are one of the most invasive plants we have, and they decimate local plant and animal diversity and thus threaten ecosystem stability and function wherever they spread. Autumn olive berries might provide cancer-fighting benefits, but so do berries of many native plants (elderberry, for example). We can take advantage of other sources of lycopene. In my view, this is a clear case where the costs of planting a nonnative species far outweigh the replaceable benefits.

Q: **To build Homegrown National Park, most of us will have to add more plants to our landscapes. But won't that increase the risk of fire in western states?**

A: What a conundrum! Climate change is causing longer and more intense droughts all over the American West, and when there is plenty of dry fuel on the ground, the risk of fire increases. It is possible, however, to design landscapes that minimize this risk. Before we started to mismanage our coniferous forests in the West by suppressing fire and allowing fuel to accumulate, low and relatively cool ground fires were frequent. They would consume accumulated grasses and shrubs every few years without jumping to the crown of large trees. We can re-create this savannah-like landscape by thinning the large trees near our homes and removing dead brush every few years, just as a ground fire would do. Fire experts suggest we clear a 100-foot circumference around the house. Another preventive tip is to be sure to remove invasive cheatgrass (*Bromus tectorum*), a winter annual from Europe that has spread throughout the West. Though green in winter and spring, it dies back to highly flammable tinder in early summer. Most of our terrible forest fires start when cheatgrass is ignited by lightning or careless people. Eradication of cheatgrass is unlikely, but you can suppress its dominance by encouraging native perennials that hold more moisture through the dry summer months and thus are less flammable.

Q: **As we take more and more resources from other creatures, why don't we see the predicted negative impact on ecosystems?**

A: We do, in fact, see these negative impacts, particularly in African and Middle Eastern ecosystems, where the impact of collapsing ecosystems can be measured in human suffering that we habitually mislabel as political unrest, famine, or failed states without acknowledging the root causes.

Unfortunately, we don't need to go to Africa to find symptoms of eroding ecosystems, because they are everywhere. Nearly one-third of the atmospheric carbon now disrupting our climate has come from removing plants around the world and releasing the carbon they once stored in the planet's soils. Flooding and wildfires are escalating (recently we got seven inches of rain in our yard from a storm that wasn't even predicted); pollinators are declining; the Aral Sea, once the earth's fourth-largest lake, has dried up—the list of extinctions is growing and the list of species now at risk of extinction is exploding. Finite freshwater stores are declining globally, there is a seasonal dead zone at the mouth of every large river in the world, ocean acidification and bleaching have destroyed thousands of miles of coral reefs, fish stocks have collapsed, micro- and macroplastic pollution is omnipresent in all oceans, lost predators have unbalanced food webs nearly everywhere—need I go on?

Q: **In *Bringing Nature Home*, you wrote that 54 percent of the United States is what you call the urban/suburban matrix. But land-use websites say only 3.5 percent of the United States is urban. How much land is really impacted by humans? I see lots of good habitat while driving around. Aren't you overstating the problem we have?**

A: Many people have questioned these statistics, because, as a nation, we want to believe that there is still a lot of nature out there. The suburban/urban matrix I refer to is the patchwork of cities and towns that now blankets most areas east of the Mississippi and more and more of the West. The matrix comprises some small, isolated patches of habitat, but it is rarely what I would consider good habitat. We worry about the destruction of the Amazon rainforest, 20 percent of which has already been logged (according to Raintree, www.rain-tree.com/facts.htm). By comparison, more than 70 percent of our Eastern forests are gone, and no one seems to notice (Brown 2006).

How did I get the 54 percent figure? I started by getting familiar with Michael Rosenzweig's work at the University of Arizona. He has spent a long and highly respected career studying the evolution and maintenance of biological diversity and is the leading authority on the matter. Rosenzweig tells us that only 5 percent of the lower forty-eight United States is relatively pristine (2003). The USDA estimates that 41 percent of the United States is in some form of agriculture, which includes monoculture tree farms that many folks mistake for wilderness. That leaves 54 percent of the United States, an area that has been chopped up into cities, suburbs, malls, roads, airports, golf courses, and other developed land, with a few habitat fragments. The reason formal definitions of how much space is urban differ from my estimates of the urban/suburban matrix is because the former are based on the density of people living in an area. Areas such as airports, malls, paved infrastructure, corporate landscapes, industrial complexes, and the like, have very few residents and thus are not defined as urban areas. In fact, such areas are often ignored altogether in land-use statistics.

Another critical part of this argument is the state of our natural areas. That patch of woods at the end of the street and the seemingly endless trees that line many of our highways look like pristine habitats to most people. But those tiny patches are unable to sustain many species precisely because they are tiny and isolated from one another. Moreover, they have been logged at least once and often multiple times over the past several centuries; the old-growth trees vital to species such as the ivory-billed woodpecker are long gone. They also have been invaded by non-native plant species that do not support our local food webs. University of Minnesota Insect Ecology Professor Richard Mack told me that cheatgrass, for example, has destroyed the sagebrush ecosystems in more than 200,000 square miles of the West. The top predators (wolves, bears, and cougars) that once controlled deer, raccoons, and possums are gone as effective controls east of the Rockies, so now populations of these prey species are far above the carrying capacity of the land to the detriment of local plant and bird communities. In my county, the carrying capacity for white-tailed deer is approximately 14 per square mile. We have 100 deer per square mile in most years, and the understory in those forests that most people think are wild places has been destroyed.

My point is simple: The most destructive part of suburbia is not the space occupied by houses and roads, although that is substantial. (In fact, the paved surface area of the four million miles of roads that crisscross the United States now covers an area over five times the size of New Jersey [Hayden 2004; Elvidge et al. 2004].) Instead, the way suburbia fragments large chunks of habitat into areas too small to sustain nature threatens our biodiversity. These are all reasons that the effects of humans on ecosystem function in the urban/suburban matrix are far greater than most people realize.

Q: What about the risks of bringing nature into our everyday spaces? Isn't interacting with nature dangerous?

A: Perhaps because of our detachment from nature, sensationalized rare events, or our innate suspicion of the natural world that once threatened us, our ability to assess risk accurately in the United States has gone haywire. A 2006–2010 government study reported that there are 88,000 deaths from alcohol every year in this country, yet one physician called West Nile virus, which killed 105 people nationwide in 2013, "a matter of life and death" and recommended that no one go outside. According to the American Cancer Society, about 9300 people are expected to die from melanoma in the United States in 2018, yet we rush to the beach every summer and to tanning salons every winter to bask in cancer-causing rays. In 2017, cities in the North sponsored large-scale, expensive mosquito fogging campaigns to combat Zika virus (campaigns that kill all of the insects they reach, not just mosquitoes), even though no Zika virus had been reported north of the Gulf Coast. And these mosquito campaigns continue, even though Zika has been totally eliminated from the United States. We have a real fear of sharks, and why not? On average, one person is killed by a shark every other year in U.S. waters. Things that we should worry about but don't include the estimated 37 children who die in hot cars each year, the 26,000 people who died in 2010 as a result of falling, and the accidental poisonings that ended the lives of 33,000 more people that year. On average, complications from obesity killed 30,000 people in 2017, almost 3500 people were killed while texting in 2015, autoerotic asphyxiation kills 600 each year, falling coconuts kill 150 people annually and lightening kills another 40–50 people each year, pet dogs killed 30 more in 2006, vending machines kill 13 each year, about 5 people die on roller coasters every year, and a whopping 300 people are killed by toasters in the United States each year. Can you guess how many people are killed by gardening with native plants?

Q: I accept that humans have changed nature in many ways, but I'm not so sure that is bad. Won't natural selection "adjust" the natural world to accommodate human changes? Why won't insects quickly adapt to novel plants and the food web will continue as it always has?

A: I wish insects would quickly adapt, but that is not how evolution works. Each year, Callery pears, oriental bittersweet, multiflora rose, Japanese honeysuckle, crown vetch, and autumn olive seed into our property from our neighbors' properties. Cindy and I continually work to remove them, but if we didn't, their numbers would increase until our property was covered in little else, exactly the way it was when we bought it. This transformation would occur in only three to five years. What would happen to the insects that depend on the oaks, red maples, tulip trees, cherries, black walnuts, and native viburnums that are crowded out in such landscapes? Would they adapt to the invasive plants in that short period of time? No, they would disappear, and so would the 897 species of caterpillars (yes, I am counting them) that depend on our native plants and that, in turn, fuel bird reproduction in our yard. Will our monarchs switch to Japanese stiltgrass when it overruns our milkweed patch? Of course not. Monarchs have been physiologically locked into eating only members of the milkweed lineage for millions of years. Evolution does not happen at the rate we are changing the food base for native animals. Unfortunately, extinction of existing fauna is the typical response to invasion by new species (Webb 2006).

Q: Most of our public lands (local parks and preserves) have been degraded by introduced plants, and land managers do not have the budget or the labor force to combat this effectively. How can we possibly manage these public lands during the age of invasive species?

A: This is a huge but not impossible challenge. One solution would be to crowd-source the management. For example, every public park or natural area is bordered by private landowners. If everyone who owned land bordering a park assumed responsibility for controlling invasive plants along their borders, the invaded area would shrink considerably. The area in which these species are controlled increases dramatically as property owners remove invasives deeper and deeper into the park. This approach would be particularly effective in linear parks (which is nearly all of our riparian park-lands). Let's say a park is 500 yards wide and 2000 yards long, and all of it has been invaded by introduced plants. That is a total area of 1 million square yards of invaded parkland. If every property owner bordering that park con-trolled invasives just 50 yards into the park along their bordering property lines, 200,000 square yards of parkland could be perpetually maintained, invasive free. If each property owner controlled invasives 100 yards into the park, nearly half of the park would be freed of invasive plants. Such prog-ress might even encourage property owners to clear invasives deeper into the park. Because plant invasions are typically most severe on park edges, this approach would address the areas most seriously impacted. Moreover, the removal of invasives would reduce the rain of seeds from introduced plants into the park, and long-term control would become easier and easier. How could we convince bordering property owners to take such action? Social pressure would do it. It is clearly a privilege to live next to a natural area. Privileges are not free; they come with strings attached in the form of responsibility for the park's well-being.

Q: Are cultivars of native plants the ecological equivalents of their straight species?

A: I get asked this question more than any other. In fact, I addressed it in *Bringing Nature Home*, but that was before research had compared eco-logical function in cultivars and their parent genotypes, so my answer was not much more than an educated guess. The interest in how well cultivars function stems from how difficult it can be to find and purchase straight species; when natives are offered in typical garden centers, the vast major-ity are cultivars that have been selected for a particular trait. Some recent studies have been designed to tackle this question, and the results suggest that the answer depends upon which trait has been selected. A study that my graduate student Emily Baisden recently finished compared insect use of straight species with use of cultivars that had been selected for six traits: changed growth habit, enhanced fruit size, enhanced fall color, disease resis-tance, leaf variegation, and leaf color changes from green to red, purple, or blue (Baisden et al. 2018). Baisden found that the only trait that consistently deterred insect herbivores was changing green leaves to red, purple, or blue. For her PhD, Annie White at the University of Vermont compared cultivars in which flower traits were changed and found that, more often than not, changing flower size, color, or shape also changed the availability and/or quality of pollen and nectar offered by the flower and thus negatively impacted pollinators.

Although some cultivar traits, such as altered growth habit and disease resistance, can make some natives good choices for residential landscapes, I have two problems with cultivars in general. First, most cultivars are propagated by cloning; this means we are planting individuals with no genetic variation at all. In the age of climate change and highly variable weather, loading our landscapes with plants that do not have the evolutionary mechanism to adapt (genetic variability) makes little sense. Second, offering only cultivars for sale perpetuates the notion that plants are simply decorations and how they interact with other species is irrelevant. I would love to see straight species sold alongside their cultivars so that people who value function over aesthetics have the option buy these plants.

Q: Doesn't this new approach to landscaping take more knowledge than most homeowners have? I don't think it will catch on because it is more complicated than sitting on a mower.

A: I have heard this concern repeatedly: homeowners don't know enough about plants or ecology to transform their traditional landscapes into living landscapes. My, my! What little faith we have. I agree that it is really easy to mow, and that increasing the number of natives in your landscape requires some knowledge about natives. But we learned how to landscape with plants that don't belong in our yards, so surly we have the intelligence to learn how to favor plants that want to be in our yards. Think about all of the new things we learn all the time. In the 1980s we learned how to program our VCRs. Nothing could be harder than that! In the 1990s we became proficient at e-mail, and today we master our smart devices in just a few days. This is complicated stuff—in my opinion, far more complicated than learning that an oak tree is a better choice than a Norway maple—yet we have measured up to the task each time new knowledge has challenged us. And if something does prove to be beyond us, we hire someone who is an expert to do the job. There is a multibillion-dollar landscaping industry ready to do whatever we ask to our properties if we are not interested in puttering ourselves. Believe me, we can do this. All we need is the motivation.

Q: Privet flowers are good sources of honey and pollen for honey bees. If we remove privet from our natural areas as part of our fight against invasive species, won't our honey bees suffer?

A: Nine species of ornamental privet have invaded more than one million acres of natural areas in the eastern United States (Hanula et al. 2009), making such areas far less natural than they used to be. One consequence of privet invasions is that what was once a diverse shrub layer comprising viburnums, blueberries, buttonbush, sheep laurel, winterberry, sweet pepperbush, mountain laurel, Virginia sweetspire, native azaleas, and myriad species of perennials and annuals is now a monoculture of privet. It is true that privet provides good forage for honey bees while it is in bloom, but that happens during one short week each year. The rest of the year it just sits there, excluding native plants that would otherwise be making pollen and nectar for our bees. Bees need a diversity of flowering plants to provide a diversity of food sources, and they need such food throughout the season. Monocultures of introduced plants such as privet, burning bush, Japanese knotweed, multiflora rose, Autumn olive, and bush honeysuckle provide just the opposite: a single species with a very short bloom period. This impacts our honey bees, to be sure, but think about what it is doing to the thousands of species of native bees that our native plant communities have relied on for pollination services for millennia. Removing invasive species from our natural areas and replacing them with a diversity of productive plants will not harm the honey bee; it will help it.

Q: Somebody told me that they want to create a bird- and insect-friendly garden, but they asked me to give them a list of nontoxic plants for kids.

A: The ironic thing is that kids don't eat plants—even the vegetables we want them to eat. They are not like horses. The first taste of a leaf that's bitter and they won't take another bite. Peaches are toxic if you eat the pit, which is full of cyanide. Same with apples, but we never eat the pit or seeds, so no one cares. Kids are already surrounded by toxic plants—every plant on the street is, in fact, toxic if you eat enough of it. Many introduced plants that we live with every day could be a problem if you ate a ton of them (such as azaleas, buttercups, lily of the valley, nandina, oleander, amaryllis, chrysanthemum, English ivy, foxglove, holly, hydrangea, and periwinkle). The real danger to kids is getting in a car. Motor vehicle accidents are the leading cause of death for children under thirteen years of age, but we manage that risk and don't worry about it. Bottom line is I don't have a list of nontoxic plants, and I also don't think you need one.

Q: **Charles Mann, who authored the 2011 book** *1493: Uncovering the New World Columbus Created,* **says that it is difficult to restore ecosystems to their natural prehuman state because humans have been manipulating ecosystems for many thousands of years. Does that mean Homegrown National Park is ill-conceived and doomed before we begin?**

A: Charles Mann is absolutely correct. It is very difficult to decide what a natural state is. But it is not difficult at all to know what a productive state is. A productive ecosystem maintains its inherent diversity and produces lots of ecosystem services. The more species in an ecosystem, the more productive it is. Scientists believe that Native Americans changed the prehuman state of North American ecosystems by hunting to extinction the large Pleistocene mammals that dominated this continent for millennia. Yet through controlled burns, they maintained the savannah-like structure of woodlands in much the same way those large mammals had done. But here's the key: although Native Americans did manipulate the plants of local ecosystems, they did not change what those plants were or how they interacted with one another. Unlike us modern Americans, Native Americans did not replace native plant communities with species from other continents, so they did not destroy the specialized coevolved relationships that glued local ecosystems together. In other words, Native Americans fiddled with but did not, on average, reduce ecosystem productivity. Besides, our goal is not to turn the clock back to a particular period and re-create ecosystems that once flourished in North America; we have already changed too many ecosystem components—from soils, to plant communities, to natural fire regimes, to top predators—for this to be possible. Instead, our goal is to restore as much ecosystem productivity as possible by reassembling the specialized relationships that encourage productivity, even if they are not exactly the same relationships that existed at a particular place 1000 or 2000 years ago.

Q: Does my yard have to be 100 percent native to join Homegrown National Park?

A: Absolutely not! There is room for compromise. Over the years, my students and I have framed our research results in terms of the ecological harm that occurs when introduced plants replace natives. With few exceptions, however, it is not the addition of introduced plants to our landscapes that destroys biodiversity, but the removal of the native plants upon which that biodiversity depends. Landscapes with a healthy dose of keystone plant genera almost always have room for some striking noninvasive introduced ornamentals without losing their ecological clout.

REFERENCES

Agrawal, A. A., J. A. Lau, and P. A. Hambäck. 2006. Community heterogeneity and the evolution of interactions between plants and insect herbivores. *The Quarterly Review of Biology* 81(4):349–76.

Annecke, D. P., and V. C. Moran. 1978. Critical reviews of biological pest control in South Africa. 2. The prickly pear, *Opuntia ficus-indica* (L.) Miller. *Journal of the Entomological Society of Southern Africa* 41(2):161–88.

Baisden, E. C., D. W. Tallamy, D. L. Narango, and E. Boyle. 2018. Do Cultivars of Native Plants Support Insect Herbivores? *HortTechnology* 28(5):596–606.

Becerra, J. X. 2007. The impact of herbivore–plant coevolution on plant community structure. *Proceedings of National Academy of Sciences USA* 104(18):7483–88.

———. 2015. On the factors that promote the diversity of herbivorous insects and plants in tropical forests. *Proceedings of National Academy of Sciences USA* 112(19):6098–6103.

Bernays, E. M., and M. Graham. 1988. On the evolution of host specificity in phytophagous arthropods. *Ecology* 69(4):886–92.

Blair, R. B. 1996. Land use and avian species diversity along an urban gradient. *Ecological Applications* 6(2):506–19.

Bormann, F. H., D. Balmori, and G. T. Geballe. 2001. *Redesigning the American Lawn: A Search for Environmental Harmony.* 2nd ed. New Haven, CT: Yale University Press.

Brewer, R. 1961. Comparative notes on the life history of the Carolina Chickadee. *The Wilson Bulletin* 73(4):348–73.

Brinkley, D. 2009. *The Wilderness Warrior: Theodore Roosevelt and the Crusade for America.* New York: Harper.

Brower, L .P., O. R. Taylor, E. H. Williams, D. A. Slayback, R. R. Zubieta, and M. I. Ramírez. 2011. Decline of monarch butterflies overwintering in Mexico: Is the migratory phenomenon at risk? *Insect Conservation and Diversity* 5(2):95–100.

Brown, W. P. 2006. *On the community composition and abundance of Delaware forest birds.* PhD Dissertation, University of Delaware.

Burgess, K. S., and B. C. Husband. 2006. Habitat differentiation and the ecological costs of hybridization: the effects of introduced mulberry (*Morus alba*) on a native congener (*M. rubra*). *Journal of Ecology* 94(6):1061–69.

Burghardt, K. T., D. W. Tallamy, and W. G. Shriver. 2009. Impact of native plants on biodiversity in suburban landscapes. *Conservation Biology* 23(1):219–24.

Burkle, L. A., J. C. Marlin, and T. M. Knight. 2013. Plant-pollinator interactions over 120 years: loss of species, co-occurrence, and function. *Science* 339(6127):1611–15.

Burton, A. 2017. Crickets in crisis. *Frontiers in Ecology and the Environment* 15:121.

Bussman, J. 1933. Experiments with the terragraph on the activities of nesting birds. *Bird-Banding* 4(1):33–40.

Cameron, S. A., J. D. Lozier, J. P. Strange, J. B. Koch, N. Cordes, L. F. Solter, and T. L. Griswold. 2011. Patterns of widespread decline in North American bumble bees. *Proceedings of the National Academy of Sciences USA* 108(2):662–67.

Carson, R. 1962. *Silent Spring.* New York: Houghton Mifflin.

Cassils, J. A. 2004. Overpopulation, sustainable development, and security. *Population and Environment* 25(3):17–94.

Chauvenet, A. L. M., and M. Barnes. 2016. Expanding protected areas is not enough. *Science* 353(6299):551–52.

Chollet, S., C. Bergman, A. J. Gaston, and J. L. Martin. 2015. Long-term consequences of invasive deer on songbird communities: Going from bad to worse? *Biological Invasions* 17(2):777–90.

Collier, M. H., J. L. Vankat, and M. R. Hughes. 2002. Diminished plant richness and abundance below *Lonicera maackii*, an invasive shrub. *The American Midland Naturalist* 147(1):60–71.

Condon, M. A., S. J. Scheffer, M. L. Lewis, and S. M. Swensen. 2008. Hidden neotropical diversity: Greater than the sum of its parts. *Science* 320(5878):928–31.

Coombs, G., and D. Gilchrist. 2018. *Native and Invasive Plants Sold by the Mid-Atlantic Nursery Industry: A Baseline for Future Comparisons.* Hockessin, DE: Mt. Cuba Center Publications.

Costello, S. L., P. D. Pratt, M. B. Rayamajhi, and T. D. Center. 1995. Arthropods associated with above-ground portions of the invasive tree, *Melaleuca quinquenervia*, in South Florida, USA. *Florida Entomologist* 86(3):300–22.

Cracknell, D., M. P. White, S. Pahl, W. J. Nichols, and M. H. Depledge. 2016. Marine Biota and psychological well-being: A preliminary examination of dose-response effects in an aquarium setting. *Environment and Behavior* 48(10):1242–69.

Cronk, L. 1999. *That Complex Whole: Culture and the Evolution of Human Behavior.* Boulder, CO: Westview Press.

Darke, R., and D. W. Tallamy. 2014. *The Living Landscape: Designing for Beauty and Biodiversity in the Home Garden.* Portland, OR: Timber Press.

Dasmann, R. F. 1968. *A Different Kind of Country.* New York: Macmillan.

Davis, M. A. 2009. *Invasion Biology.* Oxford, UK: Oxford University Press.

———. 2011. Do native birds care whether their berries are native or exotic? No. *BioScience* 61(7):501–02.

Davis, M. A., M. K. Chew, R. J. Hobbs, A. E. Lugo, J. J. Ewel, G. J. Vermeij, J. H. Brown, et al. 2011. Don't judge species on their origins. *Nature* 474:153–54.

De Kiriline Lawrence, L. 1967. A comparative life-history study of four species of woodpeckers. *Ornithological Monographs* 5:1–156.

Del Tredici, P. 2006. Brave new ecology. *Landscape Architecture* 96(2):46–52.

Diamond, J. 2011. *Collapse: How Societies Choose to Fail or Succeed*. New York: Penguin Books.

Dirzo, R., H. S. Young, M. Galetti, G. Ceballos, N. J. B. Isaac, and B. Collen. 2014. Defaunation in the Anthropocene. *Science* 345(6195):401–06.

Downey, P. O., and D. M. Richardson. 2016. Alien plant invasions and native plant extinctions: a six-threshold framework. *AoB Plants*. https://academic.oup.com/aobpla/article/doi/10.1093/aobpla/plw047/2609604. Accessed 28 November 2018.

Duncan, R. P., A. G. Boyer, and T. M. Blackburn. 2013. Magnitude and variation of prehistoric bird extinctions in the Pacific. *Proceedings of the National Academy of Sciences USA* 110(16):6436–41.

Dunn, J. L. and K. L. Garrett. 1997. *A Field Guide to Warblers of North America*. Peterson Field Guides. New York: Houghton Mifflin.

Dussourd, D. E., and T. Eisner. 1987. Vein-cutting behavior: Insect counterploy to the latex defense of plants. *Science* 237(4817):898–901.

Dyer, L. A., T. R. Walla, H. F. Greeney, J. O. Stireman III, and R. F. Hazen. 2010. Diversity of Interactions: A Metric for Studies of Biodiversity. *Biotropica* 42(3):281–89.

Eeva, T., S. Helle, J. Salminen, and H. Hakkarainen. 2010. Carotenoid composition of invertebrates consumed by two insectivorous bird species. *Journal of Chemical Ecology* 36(6):608–13.

Ehrlich, P., and A. Ehrlich. 1981. *Extinction: The Causes and Consequences of the Disappearance of Species*. New York: Random House.

Elbein, A. 2017. How Hurricane Harvey Affected Birds in Their Habitats in Texas. *Audubon*. http://www.audubon.org/news/how-hurricane-harvey-affected-birds-and-their-habitats-texas. Accessed 28 November 2018.

Elvidge, C. D., C. Milesi, J. B. Dietz, B. T. Tuttle, P. C. Sutton, R. Nemani, and J. E. Vogelmann. 2004. U.S. Constructed Area Approaches the Size of Ohio. *EOS, Transactions, American Geophysical Union* 85:233–40.

Environment Protection Agency. 2008. Sustainable Landscaping, MidAtlantic Region. http://www.epa.gov/reg3esd1/garden/presentation.htm. Accessed 14 February 2008.

Faaborg, John. 2002. *Saving Migrant Birds: Developing Strategies for the Future.* Austin, TX: University of Texas Press.

Forister, M. L., V. Novotny, A. K. Panorska, L. Baje, Y. Basset, P. T. Butterill, L. Cizek, et al. 2015. The global distribution of diet breadth in insect herbivores. *Proceedings of the National Academy of Sciences USA* 112(2):442–47.

Forseth, I. N., and A. F. Innis. 2004. Kudzu (*Pueraria montana*): History, physiology, and ecology combine to make a major ecosystem threat. *Critical Reviews in Plant Sciences* 23(5):401–13.

Fowler, J. 2016. Specialist bees of the Northeast: Host plants and habitat conservation. *Northeastern Naturalist* 23(2):305–20.

Grandez-Rios, J. M., L. L. Bergamini, W. Santos de Araújo, F. Villalobos, and M. Almeida-Neto. 2015. The effect of host-plant phylogenetic isolation on species richness, composition and specialization of insect herbivores: A comparison between native and exotic hosts. PLoS ONE 10(9):e0138031. https://doi.org/10.1371/journal.pone.0138031. Accessed 28 November 2018.

Hairston, N. G., F. E. Smith, and L. B. Slobodkin. 1960. Community structure, population control, and competition. *The American Naturalist* 94(879):421–25.

Hallmann, C.A., M. Sorg, E. Jongejans, H. Siepel, N. Hofland, H. Schwan, et al. 2017. More than 75 percent decline over 27 years in total flying insect biomass in protected areas. PLoS ONE 12(10):e0185809. https://doi.org/10.1371/journal.pone.0185809. Accessed 28 November 2018.

Hanula, J. L., S. Horn, and J. W. Taylor. 2009. Chinese privet (*Ligustrum sinense*) removal and its effect on native plant communities of riparian forests. *Invasive Plant Science and Management* 2:292–300.

Hayden, D. 2004. *A Field Guide to Sprawl.* New York: W. W. Norton.

Hobbs, R. J., et al. 2006. Novel ecosystems: Theoretical and management aspects of the new ecological world order. *Global Ecology and Biogeography* 15(1):1–7.

Hobbs, R. J., E. S. Higgs, and C. Hall. 2013. *Novel Ecosystems: Intervening in the New Ecological World Order.* Hoboken, NJ: Wiley-Blackwell.

Janzen, D. H. 1974. The deflowering of Central America. *Natural History* 83:49–53.

Jenkins, C. N., K. S. Van Houtan, S. L. Pimm, and J. O. Sexton. 2015. US protected lands mismatch biodiversity priorities. *Proceedings of National Academy of Sciences USA* 112(16):5081–86.

Jennings, V. H., and D. W. Tallamy. 2006. Composition and abundance of ground-dwelling Coleoptera in a fragmented and continuous forest. *Environmental Entomology* 35(6):1550–60.

Juniper, T. 2013. *What Has Nature Ever Done for Us? How Money Really Does Grow on Trees*. London: Profile Books.

Kardan, O., P. Gozdyra, B. Misic, F. Moola, L. J. Palmer, T. Paus, and M. G. Berman. 2015. Neighborhood greenspace and health in a large urban center. *Scientific Reports* 5 (11610). http://dx.doi.org/10.1038/srep11610. Accessed 28 November 2018.

Kaufman, S. R., and W. Kaufman. 2007. *Invasive Plants: Guide to Identification and the Impacts and Control of Common North American Species*. Mechanicsburg, PA: Stackpole Books.

Kearns, C. A., D. W. Inouye, and N. M. Waser. 1998. Endangered Mutualisms: The Conservation of Plant-Pollinator Interactions. *Annual Review of Ecology and Systematics* 29:83–112.

Knight, K. S., J. S. Kurylo, A. G. Endress, J. R. Stewart, and P. B. Reich. 2007. Ecology and ecosystem impacts of common buckthorn (*Rhamnus cathartica*): a review. *Biological Invasions* 9(8):925–37.

Knop, E., L. Zoller, R. Ryser, C. Gerpe, M. Hörler, and C. Fontaine. 2017. Artificial light as a new threat to pollination. *Nature* 548:206–09.

Kolbert, E. 2008. Turf war. *The New Yorker* 21 July.

———. 2014. *The Sixth Extinction: An Unnatural History*. New York: Henry Holt.

Kuo, F. E. (Ming). 2010. *Parks and Other Green Environments: Essential Components of a Healthy Human Habitat*. Ashburn, VA: National Recreation and Park Association.

Lande, R. 1988. Genetics and demography in biological conservation. *Science* 241(4872):1455–60.

Law, N., L. Band, and M. Grove. 2004. Nitrogen input from residential lawn care practices in suburban watersheds in Baltimore County, MD. *Journal of Environmental Planning and Management* 47(5):737–55.

Leopold, A. 1949. *A Sand County Almanac*. Oxford, UK: Oxford University Press.

Lewinsohn, T. M., V. Novotny, and Y. Basset. 2005. Insects on plants: Diversity of herbivore assemblages revisited. *Annual Review of Ecology, Evolution, and Systematics* 36:597–620.

Liddiard, R., ed. 2007. *The Medieval Park: New Perspectives*. Oxford, UK: Windgather Press.

Liebenberg, L. 2008. The relevance of persistence hunting to human evolution. *Journal of Human Evolution* 55(6):1156–59.

Louv, R. 2008. *Last Child in the Woods: Saving Our Children from Nature-Deficit Disorder*. Chapel Hill, NC: Algonquin Books.

———. 2012. *The Nature Principle: Reconnecting with Life in a Virtual Age*. Chapel Hill, NC: Algonquin Books.

———. 2016. *Vitamin N: The Essential Guide to a Nature-Rich Life*. Chapel Hill, NC: Algonquin Books.

Lovejoy, T. 1980. Preface. *Conservation Biology: An Evolutionary-Ecological Perspective*, ed. by M. E. Soulé and B. A. Wilcox. Sunderland, MA: Sinauer Associates.

MacArthur, R. 1955. Fluctuations of Animal Populations and a Measure of Community Stability. *Ecology* 36(3):533–36.

MacArthur, R. H., and E. O. Wilson. 1967. *The Theory of Island Biogeography*. Princeton, NJ: Princeton University Press.

MacIvor, J. S., and L. Packer. 2015. "Bee hotels" as tools for native pollinator conservation: A premature verdict? PLoS ONE 10(3):e0122126. https://doi.org/10.1371/journal.pone.0122126. Accessed 28 November 2018.

Mann, C. C. 2012. *1493: Uncovering the New World Columbus Created*. New York: Vintage.

Marra, P. P., and C. Santella. 2016. *Cat Wars: The Devastating Consequences of a Cuddly Killer*. Princeton, NJ: Princeton University Press.

Martin, S. G. 1971. Polygyny in the Bobolink: habitat quality and the adaptive complex. PhD Dissertation. Oregon State University.

Marzluff, J. M. 2014. *Welcome to Subirdia: Sharing Our Neighborhoods with Wrens, Robins, Woodpeckers, and Other Wildlife*. New Haven, CT: Yale University Press.

McGuire, M. 2013. The eucalypt invasion of Portugal. *The Monthly*. June 2013. https://www.themonthly.com.au/issue/2013/june/1370181600/michaela-mcguire/eucalypt-invasion-portugal. Accessed 1 December 2018.

McKinney, M. L. 2002. Urbanization, biodiversity, and conservation: The impacts of urbanization on native species are poorly studied, but educating a highly urbanized human population about these impacts can greatly improve species conservation in all ecosystems. *BioScience* 52(10):883–90.

Meine, C. 2010. *Aldo Leopold: His Life and Work*. Madison, WI: University of Wisconsin Press.

Milesi, C., S. W. Running, C. D. Elvidge, J. B. Dietz, B. T. Tuttle, and R. R. Nemani. 2005. Mapping and modeling the biogeochemical cycling of turf grasses in the United States. *Environmental Management* 36(3):426–38.

Millennium Ecosystem Assessment. 2005. *Ecosystems and Human Well-Being: Synthesis*. Washington, DC: Island Press.

Monarch Watch. 2013. Monarch Butterfly Survey Points to Lowest Numbers in 20 years. Monarch Watch Blog. https://monarchwatch.org/blog/2013/03/13/monarch-butterfly-survey-points-to-lowest-numbers-in-20-years/. Accessed 1 December 2018.

Narango, D. L., D. W. Tallamy, and P. P. Marra. 2017. Native plants improve breeding and foraging habitat for an insectivorous bird. *Biological Conservation* 213:42–50.

———. 2018. Nonnative plants reduce population growth of an insectivorous bird. *Proceedings of National Academy of Sciences USA* 115(45):11549–54.

Narango D. L., D. W. Tallamy, and K. J. Shropshire. 2018. Keystone plants are essential for insect-based food webs. In prep.

National Wildlife Foundation. 2015–2018. Native Plant Finder. https://www.nwf.org/NativePlantFinder. Accessed 29 December 2018.

Nazdrowicz, N. H., J. L. Bowman, and R. R. Roth. 2008. Population ecology of the eastern box turtle in a fragmented landscape. *Journal of Wildlife Management* 72(3):745–53.

Nickerson, C., R. Ebel, A. Borchers, and F. Carriazo. 2011. Major uses of land in the United States, 2007. Economic Information Bulletin 89. U.S. Department of Agriculture, Economic Research Service.

North American Bird Conservation Initiative. 2016. State of North America's Birds 2016. http://www.stateofthebirds.org/2016/. Accessed 1 December 2018.

Ogushi, Y., R. J. Smith, and J. C. Owen. 2017. Fruits and migrant health: Consequences of stopping over in exotic- vs. native-dominated shrublands on immune and antioxidant status of Swainson's Thrushes and Gray Catbirds. *The Condor* 119(4):800–16.

Ollerton, J., R. Winfree, and S. Tarrant, 2011. How many flowering plants are pollinated by animals? *Oikos* 120(3):321–26.

Owen, J. 2005. Farming claims almost half Earth's land, new maps show. *National Geographic News.* https://news.nationalgeographic.com/news/2005/12/1209_051209_crops_map.html. Accessed 1 December 2018.

Paine, R. T. 1969. The Pisaster-Tegula interaction: Prey patches, predator food preference, and intertidal community structure. *Ecology* 50(6):950–61.

Pappas, S. 2015. Number of Americans who don't believe in climate change rises. NBC News. https://www.nbcnews.com/science/science-news/number-americans-who-dont-believe-climate-change-rises-flna2D11943970. Accessed 21 December 2018.

Pearson, G. 2015. You're Worrying About the Wrong Bees. *Wired.* https://www.wired.com/2015/04/youre-worrying-wrong-bees/. Accessed 1 December 2018.

Peterson, R. T. 1980. *A Field Guide to the Birds.* Boston: Houghton Mifflin.

Pollan, M. 1994. Against Nativism. *The New York Times Magazine,* May 15.

Powell, K. I., J. M. Chase, and T. M. Knight. 2013. Invasive plants have scale-dependent effects on diversity by altering species-area relationships. *Science* 339(6117):316–18.

Qian, H., and R. E. Ricklefs. 2006. The role of exotic species in homogenizing the North American flora. *Ecology Letters* 9(12):1293–98.

Quammen, D. 1996. *Song of the Dodo: Island Biogeography in an Age of Extinction*. New York: Touchstone.

Rainer, T, and C. West. 2015. *Planting in a Post-Wild World: Designing Plant Communities for Resilient Landscapes*. Portland, OR: Timber Press.

Rey Banayas, J. M., A. C. Newton, A. Diaz, and J. M. Bullock. 2009. Enhancement of biodiversity and ecosystem services by ecological restoration: A meta-analysis. *Science* 325(5944):1121–24.

Richard, M., D. W. Tallamy, and A. Mitchell. 2018. Introduced plants reduce species interactions. *Biological Invasions*. https://doi.org/10.1007/s10530-018-1876-z. Accessed 21 December 2018.

Richards, O. W., and R. G. Davies. 1977. *Imms' General Textbook of Entomology*. 10th ed. *Vol. 1: Structure, Physiology, and Development*. New York: John Wiley.

Reich, P. B., D. Tilman, F. Isbell, K. Mueller, S. E. Hobbie, and D. F. B. Flynn. 2012. Impacts of biodiversity loss escalate through time as redundancy fades. *Science* 336(6081):589–92.

Rosenthal, G. A., and M. R. Berenbaum, eds. 2012. *Herbivores: Their Interactions with Secondary Plant Metabolites*. 2nd ed. New York: Academic Press.

Rosenthal, G. A., and D. H. Janzen, eds. 1979. *Herbivores: Their Interaction with Secondary Plant Metabolites*. New York: Academic Press.

Rosenzweig, M. L. 2003. *Win-Win Ecology: How the Earth's Species Can Survive in the Midst of Human Enterprise*. New York: Oxford University Press.

Sandom, C., S. Faurby, B. Sandel, and J. Svenning. 2014. Global late Quaternary megafauna extinctions linked to humans, not climate change. *Proceedings of the Royal Society B* 281:20133254. http://dx.doi.org/10.1098/rspb.2013.3254. Accessed 1 December 2018.

Saunders, D. A., R. J. Hobbs, and C. R. Margules. 1991. Biological consequences of ecosystem fragmentation: a review. *Conservation Biology* 5(1):18–32.

Schmitz, O. J., P. A. Hambäck, and A. P. Beckerman. 2000. Trophic cascades in terrestrial systems: A review of the effects of carnivore removals on plants. *American Naturalist* 155(2):141–53.

Schueler, T. R. 2010. The Clipping Point: Turf Cover Estimates for the Chesapeake Bay Watershed and Management Implications. Chesapeake Stormwater Network Technical Bulletin No. 8.

Smith, S. B., S. A. DeSando, and T. Pagano. 2013. The value of native and invasive fruit-bearing shrubs for migrating birds. *Northeastern Naturalist* 20(1):171–84.

Smith, S. B., A. C. Miller, C. R. Merchant, and A. F. Sankoh. 2015. Local site variation in stopover physiology of migrating songbirds near the south shore of Lake Ontario is linked to fruit availability and quality. *Conservation Physiology* 3(1):cov036.

Southwood, T. R. E. 1972. "The insect/plant relationship: an evolutionary perspective," in *Insect-Plant Relationships*. Hoboken, NJ: Blackwell Scientific Publications.

Stewart, R. M. 1973. Breeding Behavior and Life History of the Wilson's Warbler. *The Wilson Bulletin* 85(1):21–30.

Stolzenburg, W. 2008. *Where the Wild Things Were: Life, Death, and Ecological Wreckage in a Land of Vanishing Predators*. New York: Bloomsbury.

———. 2011. *Rat Island: Predators in Paradise and the World's Greatest Wildlife Rescue*. New York: Bloomsbury.

Stranahan, S. Q. 1993. *Susquehanna: River of Dreams*. Baltimore: Johns Hopkins University Press.

Strong, D. R., J. H. Lawton, and R. Southwood. 1984. *Insects on Plants: Community Patterns and Mechanisms*. Cambridge, UK: Harvard University Press.

Stutchbury, B. 2007. *Silence of the Songbirds*. New York: Walker.

Stutchbury, B. J., J. M. Rhymer, and E. S. Morton. 1994. Extrapair paternity in hooded warblers. *Behavioral Ecology* 5(4):384–92.

Tallamy, D. W. 2004. Do alien plants reduce insect biomass? *Conservation Biology* 18(6):1689–92.

———. 2007. *Bringing Nature Home: How You Can Sustain Wildlife with Native Plants.* Portland, OR: Timber Press.

Tallamy, D. W., and K. J. Shropshire. 2009. Ranking lepidopteran use of native versus introduced plants. *Conservation Biology* 23(4):941–47.

Tewksbury, L., R. Casagrande, B. Blossey, P. Häfliger, and M. Schwarzländer. 2002. Potential for biological control of *Phragmites australis* in North America. *Biological Control* 23(2):191–212.

Tolmé, P. 2017. The U.S. biodiversity crisis. National Wildlife Federation. https://www.nwf.org/Magazines/National-Wildlife/2017/Feb-March/ Conservation/Biodiversity. Accessed 1 December 2018.

Tymkiw, E. L., J. L. Bowman, and W. G. Shriver. 2013. The effect of white-tailed deer density on breeding birds in Delaware. *Wildlife Society Bulletin* 37(4):714–24.

United States Bureau of the Census. 1991. Statistical Abstract of the United States. Washington, DC. https://www.census.gov/library/publications/1991/ compendia/statab/111ed.html. Accessed 18 December 2018.

United States Department of Agriculture. 2006. Maryland turfgrass study 2005. https://www.nass.usda.gov/Statistics_by_State/Maryland/Publications/ Miscellaneous/turfgrass2006.pdf. Accessed 1 December 2018.

United States Fish & Wildlife Service. 2000. Homeowner's Guide to Protecting Frogs - Lawn & Garden Care. U. S. Fish & Wildlife Service, Division of Environmental Contaminants, Arlington, Virginia.

Van Dyke, F. 2008. "The History and Distinctions of Conservation Biology," in *Conservation Biology.* Dordrecht, the Netherlands: Springer.

Wagner, D. L. 2005. *Caterpillars of Eastern North America: A Guide to Identification and Natural History.* Princeton, NJ: Princeton University Press.

Wagner, D. L., D. F. Schweitzer, J. B. Sullivan, and R. C. Reardon. 2012. *Owlet Caterpillars of Eastern North America.* Princeton, NJ: Princeton University Press.

Walker, B. H. 1992. Biodiversity and ecological redundancy. *Conservation Biology* 6(1):18–23.

Wang, D., and T. MacMillan. 2013. The benefits of gardening for older adults: A systematic review of the literature. *Activities, Adaptation and Aging* 37(2):153–81.

Wargo, J., N. Alderman, and L. Wargo. 2003. *Risks from Lawn-Care Pesticides*. Environment & Human Health, Inc., North Haven, CT. http://www.ehhi.org/lawnpest_full.pdf. Accessed 1 December 2018.

Webb, S. D. 2006. The Great American Biotic Interchange: Patterns and Processes. *Annals of the Missouri Botanical Garden* 93(2):245–57.

Weis, R. 2009. National Insect Killing Week. *Decorah Journal*. July 16, B3.

Werdelin, L. 2013. Early humans, not climate change, decimated Africa's large carnivores. *Scientific American* 309(5).

Wilcove, D. S. 2008. *No Way Home: The Decline of the World's Great Animal Migration*. Washington, DC: Island Press.

Williams, P., and S. Jepsen, eds. 2014. IUCN Bumblebee Specialist Group Report 2014. http://www.xerces.org/wp-content/uploads/2015/03/2014-bbsg-annual-report.pdf. Accessed 1 December 2018.

Wilson, E. O. 1987. The little things that run the world (The importance and conservation of invertebrates). *Conservation Biology* 1(4):344–46.

———. 1992. *The Diversity of Life*. Cambridge, UK: Belknap Press.

———. 1994. *Naturalist*. Washington, DC: Island Press.

———. 2017. *Half-Earth: Our Planet's Fight for Life*. New York: Liveright.

Wolf, K. L. 2014. Greening the City for Health. *Communities & Banking* 25(1):10–12.

World Bank Group. 2018. Agricultural land data. http://data.worldbank.org/indicator/AG.LND.AGRI.ZS. Accessed 9 January 2019.

Zaya, D. N., S. A. Leicht-Young, N. B. Pavlovic, K. A. Feldheim, and M. V. Ashley. 2015. Genetic characterization of hybridization between native and invasive bittersweet vines (*Celastrus* spp.). *Biological Invasions* 17(10):2975.

INDEX

The BESTSELLING CLASSIC that will CHANGE THE WAY YOU SEE YOUR GARDEN!

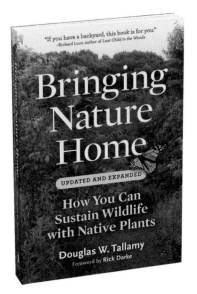

"If you have a backyard, this book is for you."
—Richard Louv, author of Last Child in the Woods

Bringing Nature Home

UPDATED AND EXPANDED

How You Can Sustain Wildlife with Native Plants

Douglas W. Tallamy

Foreword by Rick Darke

PRAISE AND ACCLAIM FOR
Bringing Nature Home

"Doug Tallamy's message is surprisingly simple—if you have enough space to grow a tree or shrub, you can do your part to help stop the extinction crisis. This book provides help to anyone who wants to become part of the solution: just get outdoors and pay attention to the living things around you."

—Richard Louv, author of *Last Child in the Woods*

"If you cut down the goldenrod, the wild black cherry, the milkweed and other natives, you eliminate the larvae, and starve the birds. This simple revelation about the food web—and it is an intricate web, not a chain—is the driving force in *Bringing Nature Home*."

—The New York Times

"Tallamy imparts an encouraging message: it's not too late to save the ecosystem-sustaining matrix of insects and animals, and the solution is as easy as replacing alien plants with natives." —*Booklist*

"Provides the rationale behind the use of native plants, a concept that has rapidly been gaining momentum.... Makes a case for native plants and animals in a compelling and complete fashion."

—The Washington Post

"A compelling argument for the use of native plants in gardens and landscapes."

—Landscape Architecture Magazine

"An essential guide for anyone interested in increasing biodiversity in the garden."

—The American Gardener

"Will persuade all of us to take a look at what is in our own yards with an eye to how we, too, can make a difference. It has already changed me." —*Traverse City Record-Eagle*

"Delivers an important message for all gardeners: choosing native plants fortifies birds and other wildlife and protects them from extinction." —*WildBird Magazine*

"This is the 'it' book in certain gardening circles. It's really struck a nerve."

—Philadelphia Inquirer

"An informative and engaging account of the ecological interactions between plants and wildlife, this fascinating handbook explains why exotic plants can hinder and confuse native creatures, from birds and bees to larger fauna."

—USA Today